Energy Efficiency Improvements with Emission Abatement for Energy Sustainability

K. J. Sreekanth
Editor

Energy Efficiency Improvements with Emission Abatement for Energy Sustainability

Energy Statistics

Springer

Editor
K. J. Sreekanth
Renewable Energy and Energy Efficiency Program (REEEP)
Energy and Building Research Center
Kuwait Institute for Scientific Research (KISR)
Kuwait City, Kuwait

ISBN 978-981-97-9479-9 ISBN 978-981-97-9480-5 (eBook)
https://doi.org/10.1007/978-981-97-9480-5

This Springer imprint is published by the registered company Springer Nature Singapore Pte Ltd.
The registered company address is: 152 Beach Road, #21-01/04 Gateway East, Singapore 189721, Singapore

Preface I

In an era defined by the urgent need for sustainable solutions to address global challenges, the pursuit of energy efficiency and emission abatement stands at the forefront of our collective responsibility. As we navigate the complex landscape of environmental stewardship, the significance of advancements in Energy Efficiency Improvements with Emission Abatement for Energy Sustainability cannot be overstated.

This comprehensive volume, meticulously crafted by a team of experts, researchers, and thought leaders, underlines the critical intersection between energy efficiency, emission reduction, and the overarching goal of achieving a sustainable energy future. As our world grapples with the impacts of climate change and the finite nature of traditional energy resources, the imperative to innovate and implement transformative measures becomes increasingly apparent.

Springer Nature's commitment to fostering knowledge dissemination and promoting interdisciplinary collaboration is exemplified in this timely and insightful collection of contributions. The book traverses the intricate realms of technology, policy, and practice, offering a holistic perspective on the multifaceted challenges associated with energy efficiency improvements and emission abatement.

The chapters within this volume span a spectrum of topics, ranging from cutting-edge technologies and engineering solutions to policy frameworks and socioeconomic considerations. By synthesizing diverse perspectives, the book provides a valuable resource for researchers, practitioners, policymakers, and students alike, fostering a deeper understanding of the intricate dynamics involved in achieving energy sustainability.

In the pursuit of a sustainable energy future, the synergy between energy efficiency improvements and emission abatement emerges as a linchpin for transformative change. The book serves as a guiding compass, navigating through the complexities of our energy landscape and illuminating pathways toward a more sustainable, resilient, and low-emission future.

As we embark on this journey toward energy sustainability, may this book inspire, inform, and catalyze the collaborative efforts needed to usher in an era where energy efficiency and emission abatement are not merely aspirations but integral components of our shared commitment to safeguarding the planet for future generations.

Dr. K. J. Sreekanth
Research Scientist
Energy and Building Research Center
Kuwait Institute for Scientific Research
(KISR)
Kuwait City, Kuwait

Preface II

In the current epoch, defined by rapid technological advancements and burgeoning global challenges, the discourse on sustainable development has become paramount. Central to this dialogue is the energy sector, a linchpin of modern society, influencing economic vitality, environmental integrity, and social well-being. As we stand at the intersection of innovation and responsibility, the imperative to re-evaluate our energy practices has never been more pressing. This book, *Energy Efficiency Improvements with Emission Abatement for Energy Sustainability*, emerges as a timely and comprehensive exploration into the multifaceted realms of energy efficiency and emission abatement. As the global community grapples with the consequences of climate change and seeks a harmonious balance between energy demands and ecological equilibrium, this compilation aspires to be both a guidebook and a catalyst for transformative action.

The urgency of addressing energy sustainability cannot be overstated. The prevailing models of energy consumption, characterized by a heavy reliance on fossil fuels, have not only driven economic growth but have also precipitated environmental degradation and climate change. The dual challenge of ensuring a consistent energy supply while mitigating adverse impacts on the planet requires a paradigm shift in our approach. This book delves into the vital intersection of energy efficiency improvements and emission abatement strategies. It seeks to bridge the gap between theory and practice, offering insights into the latest research, innovative technologies, and practical solutions that can usher in a new era of sustainable energy use. The landscape of energy sustainability is vast and complex. It encompasses technological innovations, policy frameworks, societal behaviors, and the interplay of market forces. Recognizing this complexity, the chapters in this book traverse a diverse array of topics, providing readers with a nuanced understanding of the challenges and opportunities inherent in the pursuit of sustainable energy.

The journey begins with an exploration of the foundations of energy statistics and energy efficiency in the first two chapters "Introduction to Energy Statistics" and "Importance of Energy Balance in Energy Statistics and Energy Policies". From the fundamentals of energy statistics to the role of advanced technologies, readers will gain a comprehensive grasp of the principles that underpin the quest for greater energy efficiency and energy balance in energy statistics. As we delve into the various sections, these chapters illuminate the specific challenges and tailored solutions relevant to each domain.

The subsequent two chapters "Energy Efficiency and Emission Abatement—an Indian Perspective" and "Emission Abatement: A Cop28 Outlook for Countries" unfold the intricacies of emission abatement strategies, providing specific focus on COP27 and COP28. The possibility of carbon capture and storage, emission mitigation, and the impact of research on climate change takes center stage. Through a synthesis of case studies and empirical research, the book demystifies the practical implementation of these strategies, showcasing real-world examples of success stories and lessons learned.

In the complex tapestry of global energy dynamics, oil-rich countries occupy a unique and consequential position. Their vast energy reserves not only propel their economic engines but also play a pivotal role in shaping the global energy landscape. Within this context, understanding the intricate relationship between energy statistics and energy consumption is paramount. Chapter "Energy Statistics Influence on Energy Consumption for Oil-Rich Countries" delves into the nuanced interplay, offering insights into how statistical analyses can illuminate, guide, and potentially transform the energy trajectories of these nations. This chapter endeavors to unravel the layers of influence that energy statistics wield in oil-rich countries, dissecting their impact on policy formulation, economic development, and environmental sustainability.

The sixth chapter "Drivers of Human Behavior to Meet UN Sustainable Development Goals (SDGs) in Relation With Energy" embarks on a profound journey into the sophisticated interplay between individual and collective actions, societal norms, and the quest for sustainable energy. It is a narrative that transcends disciplinary boundaries, weaving together insights from psychology, sociology, economics, and environmental science to unravel the complexities that underpin the human-energy nexus.

The foundation of this exploration rests on the understanding that the transition to sustainable energy is not solely a technological challenge but a deeply human one. As we confront the realities of climate change and resource depletion, the choices we make as individuals, communities, and societies become catalysts for transformative change. Through the lens of behavioral science, this chapter delves into the drivers that propel or impede our collective journey towards a more sustainable energy future. As we stand on the cusp of a transformative era, may this exploration into the drivers of human behavior in relation to energy inspire a collective awakening.

Recognizing that the judicious use of energy is not only a catalyst for economic growth but also a linchpin for achieving the United Nations Sustainable Development Goals (SDGs), the last chapter "Energy Policy Analyses, Energy Transition and Sustainability" unfolds as a comprehensive exploration into the dynamic realm of energy-efficient policies, with a suitable example. This compilation in this chapter is not confined to theoretical musings but is anchored in the practicalities of policy implementation. It elucidates the complexities of aligning national policies with global sustainability objectives, navigating the intricacies of regulatory frameworks, and mitigating potential challenges.

One of the distinguishing features of this book is its commitment to a holistic approach. Energy sustainability is not a monolithic challenge; it is a confluence of technological innovation, policy evolution, and societal transformation. As such, the chapters are curated to reflect the interdisciplinary nature of the field, drawing on insights from engineering, economics, environmental science, and policy analysis. Moreover, the book emphasizes the importance of international collaboration. In a world where borders are increasingly porous in the face of global challenges, the pursuit of sustainable energy necessitates a collective effort. By examining successful cross-border initiatives and policy frameworks, the book inspires readers to envision a future where nations collaborate to address shared energy challenges.

While the book serves as a compendium of knowledge, it is not meant to be a passive repository of information. Rather, it is a call to action. Each chapter concludes with practical recommendations and actionable insights, empowering readers to translate knowledge into tangible outcomes. Whether you are an academic researcher, a policymaker, an industry professional, or a concerned citizen, this book endeavors to equip you with the tools to contribute meaningfully to the cause of energy sustainability. The transition to a sustainable energy future is not without its obstacles. Economic considerations, political dynamics, and the inertia of established systems pose formidable challenges. However, the book contends that these challenges are not insurmountable. Through innovation, collaboration, and a shared commitment to a sustainable future, we can navigate the complexities of the energy landscape and forge a path towards a more resilient and equitable world.

Let us follow a spirit of curiosity, inquiry, and optimism as we travel through the pages of "Energy Efficiency Improvements with Emission Abatement for Energy Sustainability." The challenges we face are formidable, but so are the opportunities for positive change. By arming ourselves with knowledge, embracing innovation, and fostering collaboration, we can collectively shape a future where energy sustains not only our economies but also the intricate web of life on our planet.

In the words of Margaret Mead, "Never doubt that a small group of thoughtful, committed citizens can change the world; indeed, it's the only thing that ever has." May this book inspire and empower the readers to be part of that small group, catalyzing transformative change for the benefit of present and future generations. Together, let us forge a path towards a more sustainable and equitable world, where the principles of the United Nations Sustainable Development Goals are not merely aspirations but the compass guiding our shared journey.

The creation of this book has been a collaborative endeavor, bringing together the expertise of researchers, practitioners, and academicians from around the globe. Their dedication to advancing the discourse on energy sustainability is evident in the depth and breadth of the content presented. I extend my sincere gratitude to all the contributors for their invaluable insights and commitment to the shared goal of a sustainable energy future. As editor of this book, I would like to acknowledge all the authors for their valuable contributions. Also, the reviewers who took part in the review process deserve special appreciation. Without their support, this book would not have become a reality. Furthermore, I wish to acknowledge the valuable contribution and support of my colleague, Ms. Aarthi Elumalai, for her help in structuring and formatting the chapters.

Lastly, I owe an enormous debt of gratitude to Springer Nature Singapore Pte Ltd., Singapore, for publishing this book. This book wouldn't have been possible without the wholehearted support and association from them, who always provided unrelenting inspiration and guidance.

Let the exploration begin.

Kuwait City, Kuwait
December 2023

K. J. Sreekanth

Contents

Introduction to Energy Statistics

Saad Alshahrani, Ali Ahmed Alqahtani, and C. Ahamed Saleel

Abstract This chapter comprehensively analyzes energy statistics from a country's perspective. It emphasizes the importance of energy data in shaping policies, managing resources, promoting sustainability, and economic planning. It includes detailed tables and figures, covering energy indicators, aggregators, sustainable development goals, energy-economy analysis, and the environmental impact of energy transactions. It also focuses on energy consumption patterns, efficiency, renewable energy sources, and their environmental implications, to aid decision-making and promote sustainable energy management. The chapter emphasizes the crucial role of energy statistics in managing national energy resources, formulating governmental policies, and achieving sustainability goals. It highlights that accurate and up-to-date energy statistics are essential for envisioning a secure and sustainable future. The document underscores the importance of these statistics in decision-making processes, benefiting both the nation and its citizens, and guiding towards a more environmentally responsible and energy-efficient future.

Keywords Energy statistics · Resource management · Sustainability · Energy aggregators · Sustainable Development Goals (SDG) · Environmental impact

S. Alshahrani · A. A. Alqahtani · C. A. Saleel (✉)
Department of Mechanical Engineering, College of Engineering, King Khalid University, Asir-Abha 61421, Kingdom of Saudi Arabia
e-mail: ahamedsaleel@gmail.com; aveetil@kku.edu.sa

S. Alshahrani
e-mail: saadayed@kku.edu.sa

A. A. Alqahtani
e-mail: alahalqahtani@kku.edu.sa

K. J. Sreekanth (ed.), *Energy Efficiency Improvements with Emission Abatement for Energy Sustainability*, https://doi.org/10.1007/978-981-97-9480-5_1

Introduction

Energy consumption patterns and average energy use performance depend on the country's energy resources. Insights from Energy statistics are important for decision-making, making important policies, and improving the energy industry. This chapter mainly discusses energy aggregators, indicators, and Sustainable development goals in energy versus economy analysis. Energy-related things which depend on environmental significance [1]. Below is a table of Country Energy statistics and their Significance, key elements, and challenges. These can be observed in Table 1.

We can conclude that the energy statistics of the entire nation help manage energy resources, form governmental policies and fulfil sustainability goals. The government can make decisions according to the energy statistics that benefit the citizens. The accurate and current energy statistics information helps to make a vision about a secure and sustainable future [2].

Energy Aggregators

Energy aggregators, as shown in Fig. 1, are systems that systematize and control energy from several sources. These are commonly called energy aggregators or electricity registers. This can be power providers, customers, utilities, and other established players. Electricity registers have a vital role in speeding up collecting records and analyzing them to help the running of an electricity market and make energy more efficient [3, 4].

Energy aggregators perform various functions and tasks, which include:

i. Data Collection: Energy registers collect data from power generators, renewable energy installations, industrial facilities, commercial buildings and homes. The data include consumption, generated and grid-related information.
ii. Data Aggregation: After collecting the whole data, it was combined into one consistent format. This is much easier to analyze because it is an organized document.
iii. Data Management: Aggregators need to handle and secure the data. This is done by creating database management systems. It will go through validation, quality control, and security procedures.
iv. Data Analysis: Consumers can understand the gathered data using data analytics they are able find patterns, trends, anomalies and optimization of energy use.
v. Market Participation: Aggregators play an important role in the energy market to connect small-scale energy producers (such as solar panel owners) with the grid or market operator. They combine energy production from various dispersed sources, demand response of energy, and energy trading for their customers.

Table 1 Energy statistics—an overview

Significance	Key elements	Challenges
Policy formulation: Vital to formulating a nation's energy policies and strategies	Energy production: Fetches information about the extraction, generation, and imports and exports of many types of energy	Quality of the data: Ensure the accuracy and dependability of energy data
Resource management: To maximize resource use and revenue generation of a country	Energy consumption: Helps to understand how energy is used and ways of increasing energy efficiency	Timeliness: Timely data collection helps in effective decision-making and the development of strong energy policies
Sustainability: To achieve environmental sustainability objectives	Energy trade: Assess the trade balance and analyze energy security benefits	International comparability: Facilitate meaningful comparisons between nations and promote energy cooperation
Economic planning: Predicting patterns in energy supply and demand	Renewable energy: Generation and consumption patterns of various renewable energy sources help to comprehend the sustainability targets	Data adccessibility: Making energy data available to decision-makers, academics, and the general public
Development of infrastructure: For planning energy infrastructure, such as power plants, pipelines, and transmission networks	Emissions data: Climate change mitigation initiatives	Changing energy landscape: Energy statistics methodology and classifications need to be updated frequently due to the rapid growth of technology and changes in energy sources
	Energy prices: Energy costs, Affordability, Market regulation, taxation, and subsidy strategies	
	Energy accessibility: Monitoring the availability of cutting-edge and dependable energy services in urban and rural areas aids in addressing energy poverty	

vi. Demand Response: Group of customers or distributed energy resources (DERs) for demand response programs, Aggregators help utilities and grid operators in managing power demand. During high demand managed by load reduction or shifts, that helps grid stabilization and reduces the electricity prices.

vii. Energy Efficiency Programs: Using Data analysis aggregators help both companies and individuals with energy efficiency improvements, and it implements energy saving measures

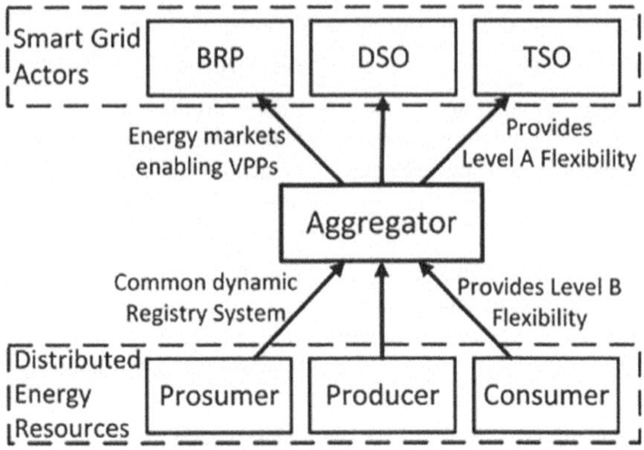

Fig. 1 Energy aggregators' systems [5]

viii. Billing and Settlement: Energy usage and transactions of many parties' precise tracking help energy aggregators order billing and payment procedures. This is particularly useful in detailed multi-party energy agreements.
ix. Compliance and Reporting: By keeping detailed and auditable records of all energy-related operations, energy aggregators facilitate with compliance with energy rules and reporting requirements.
x. Market Access: Smaller, decentralized energy producers (microgrids or community solar projects) can participate in energy markets with the help of aggregators, gaining access to revenue flow they might not be able to access on their own.

Energy Aggregators are important for energy system optimization, support energy efficiency, and consist of energy sources that enhance overall sustainability and reliability of energy supply with rising decentralization and making use of clean energy technology. Their service is really helpful for the fast-changing energy landscape.

Energy Indicators

Energy indicators are numerical standards of measurement that are used to calculate different aspects of a country's energy sector. Energy indicators provide valuable information about a country's ability to produce, use, and sustainably use energy. These indicators provide trend analysis and serious monitoring for the energy sector as well as policy development and decision-making [6]. The list of typical energy indicators is as follows:

1. Energy Production: These standards of measurement calculate the quantity of energy produced inside a country. They consist of:

    - Total Primary Energy Production: Total Energy generated in a country includes renewable and fossil fuel resources.

- Energy Production by Source: Total Energy generated in a country from various sources includes fossil fuel and renewable resources (wind, sun, hydro, biomass).

2. Energy Consumption: The energy consumption indicator evaluates the nation's overall usage of energy, in various industries including transportation, Commercial, residential and business. Important signs include

   - Total Primary Energy Consumption: The total amount of energy consumed inside a country.
   - Per Capita Energy Consumption: Energy consumption per individual, This is frequently used for calculate energy efficiency between different nations.
   - Energy Intensity: An energy efficiency measurement that is frequently expressed as energy usage per unit of GDP (gross domestic product).
   - Energy Trade: This measurement of a nation's energy imports and exports can have an important effect on a nation's economy and national security. Important points include:
   - Net Energy Imports: A nation's energy trade balance is represented as the difference between energy imports and exports.

3. Energy Import Dependency: The Proportion of imported energy used which may indicate concerns about energy security

   - Renewable Energy Indicators: The usage of renewable energy sources which is important for sustainability, and reduce the greenhouse gas emissions is monitored by these measurements.
   - Renewable Energy Share: The proportion of renewable energy sources in total energy consumption.
   - Installed Renewable Energy Capacity: the total installed capacity of renewable energy sources (such as wind, solar, and hydro) in a country. Renewable energy statistics of 2022, revealed by a new global status report calculate the rapid expansion of renewable energy sources, the change of the energy system is slow, and fossil fuel is still dominating the main source world's energy system. Global de-carbonization will remain a dream without proper political intervention. The details of Statistics on renewable energy can be seen in Fig. 2.

4. Energy Efficiency Indicators: These measurements reflect a nation's capacity to generate goods and services with little energy input. Examples include:

   - Energy Efficiency Ratio: The proportion of energy input to useful energy production in a particular process or industry.
   - Energy Savings: Measurable reductions in energy utilization cause energy efficiency techniques.

5. Carbon Emissions: This are all the measure of a country's greenhouse gas emissions from both the production and usage of energy. Important indicators Comprises:

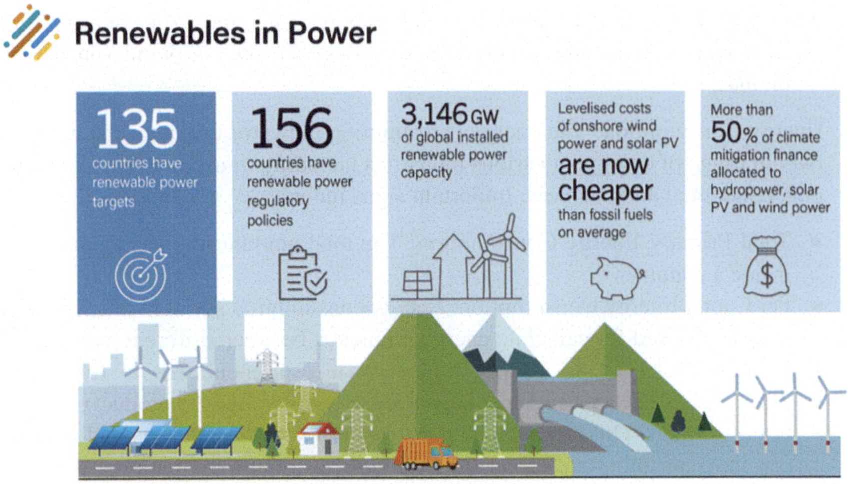

Fig. 2 Detailed statistics on renewable energy [7]

- CO_2 Emissions from Energy Use: Emissions of carbon dioxide followed by energy-related Process.
- Carbon Intensity: The measure of how much amount of carbon dioxide is emitted or absorbed per unit of energy.

6. Energy Access Indicators: The focus of these measurements is on the percentage of a population that has access to dependable and modern energy services, such as electricity and facilities for healthy cooking.
7. Energy Prices: indicators of the cost of energy, such as:
   - Average Energy Prices: Energy's cost per unit, such as per kWh or per gallon of petrol.
   - Price Trends: Price variations for energy over time.
8. Energy Reserves: Indications of a country's fossil resource reserves, such as coal, natural gas, and oil, in terms of their size.
9. Energy Investment: Metrics used to assess governmental and private sector energy sector investments in Research and development, infrastructure, and other areas.

Energy indicators are essential tools for identifying a country's energy requirement setting energy goals, and developing policies to ensure a sustainable, secure and productive energy future. These are helping in conducting cross-national comparisons as well as measuring progress towards global energy and environmental goals.

Sustainable Development Goals

The Sustainable Development Goals (SDG), also called global goals, is a set of 17 comprehensive goals approved by UN members in September 2015. It is developed as a part of the 2030 agenda for sustainable development, a global action plan to address mainly social, economic, and environmental issues the world is presently experiencing. Compared to the Millennium Development Goals (MDG), the Sustainable Development Goals (SDGs) provide a more detailed and ambitious framework to promote sustainable development [8].

The primary goal of the SDGs is to save the environment while promoting prosperity for All Several development-related concerns are solved by these goals, including the eradication of poverty, the growth of the economy, social inclusion, environmental sustainability, and peace and justice. Each goal has a specific set of indicators and milestones to monitor progress. The 17 sustainable development Goals are as follows:

1. No Poverty: All types of poverty must end.
2. Zero Hunger: Eradicate hunger, establish greater nutrition and food security, and encourage environmentally friendly farming.
3. Good Health and Well-being: Encourage and guarantee healthy lives for people of all ages.
4. Quality Education: Ensure inclusive and equitable quality education and provide chances for everyone to learn new things throughout their lives.
5. Gender Equality: Bring about gender parity and empower all women and girls.
6. Clean Water and Sanitation: Make sure that everyone has access to clean water and sanitation that is managed sustainably.
7. Accessible and Sustainable Energy: Ensure that everyone has access to modern, sustainable, inexpensive energy.
8. Decent Work and Economic Growth: Promote full and productive employment, decent labor for everyone, and sustained, inclusive, and sustainable economic growth.
9. Industry, Innovation, and Infrastructure: Foster innovation, develop inclusive and sustainable industry, and build resilient infrastructure.
10. Reduced Inequality: Reduce inequality both within and across nations.
11. Sustainable Cities and Communities: Ensure inclusiveness, safety, resilience, and sustainability in urban areas and human settlements.
12. Responsible Consumption and Production: Ensure sustainable patterns of production and consumption.
13. Climate Action: Adopt immediate steps to mitigate the effects of climate change.
14. Life Below Water: Protect the oceans, seas, and marine resources and utilize them wisely for sustainable development.
15. Life on Land: Stop biodiversity loss, protect and restore terrestrial ecosystems, manage forests sustainably, fight desertification, and prevent and reverse land degradation.

16. Peace, Justice, and Strong Institutions: Encourage inclusive and peaceful societies for sustainable development, ensure that everyone has access to justice, and create inclusive, effective, and accountable institutions at all levels.
17. Partnerships for the Goals: Boost the global partnership for sustainable development and strengthen its implementation strategies.

Solving one aspect of sustainable development frequently supports advancement in other areas these goals are interconnected and mutually beneficial. The SDGs aim to mobilize financial resources, promote international cooperation, and guide regional and global development activities to make the world more just wealthy, and sustainable by the year 2030. The various parties involved in implementing the SDGs and tracking their success include government civil organizations, Companies, and people. Timely Data is available in the link at https://sdgstoday.org/ [9].

Energy-Economy Analysis

A detailed analysis and assessment of the relationships between the energy sector and other economic sectors on a national scale includes an energy economy study. This kind of study is needed to completely understand the effects of energy production, consumption, and policy on social well-being, employment, environmental sustainability and economic excellence [10]. From a national point of view, the essential features and elements of examination of energy economy are as follows:

i. Energy Supply and Demand Analysis:

- Energy Production: Analyze the country's energy output, including nuclear energy, imports, and exports of fossil fuels (such as coal, oil, and natural gas), and renewables (like wind, solar, and hydropower).
- Energy Consumption: Look at how energy is used in several sectors, including transportation, housing, and commerce. Establish the trends, patterns, and causes of energy consumption.

ii. Economic Impact Assessment:

- Gross Domestic Product (GDP): Analyze the relationship between energy consumption and GDP to look for links between energy use and economic expansion.
- Energy Intensity: A measure of the quantity of energy required to produce one unit of economic output, such as GDP, is the energy intensity.
- Employment: Analyze the number of employment that is generated by the energy sector, both directly (such as during the generation of energy) and indirectly (such as in related industries).

iii. Energy Efficiency Analysis:

- By examining energy consumption per unit of economic production, also known as energy intensity or energy productivity, one can evaluate the performance of the nation's energy efficiency.
- Determine how industrial processes, transportation, and building systems may all be made more energy efficient.

iv. Environmental Impact Assessment:

- Research the environmental impacts of energy consumption and production, including greenhouse gas emissions, air and water pollution, and resource depletion.
- To decrease the impact on the environment, consider the potential of switching to greener energy sources and technologies.

v. Energy Policy Evaluation:

- Analyze how effectively present energy laws and regulations further both economic and energy goals.
- Analyze how government incentives and subsidies affect energy production, consumption, and investment.

vi. Scenario Modeling:

- To analyze the potential financial impacts of various energy pathways and policy options, create scenarios. Think about the financial impacts of, say, implementing carbon pricing systems or transitioning to renewable energy sources.

vii. Energy Security and Trade:

- Determine the implications on energy security and trade balances by analyzing the dynamics of the country's energy import and export.
- Analyze potential energy supply disruption techniques and weaknesses.

viii. Investment and Infrastructure Analysis:

- Determine whether investments in energy infrastructure, such as power plants, transmission grids, and transportation networks, are necessary.
- Calculate the economic benefits of improved infrastructure, such as higher productivity and more jobs.

ix. Social and Equity Considerations:

- Consider how social and equity issues, such as how distribution is affected by energy pricing and whether underprivileged groups can obtain energy services, are impacted by energy policies.

x. International and Regional Context:

- Examine how the nation's energy-economic interactions fit into the larger global and regional energy picture, taking into account the effects on trade ties and international climate commitments.

Energy-economy study from a national point of view can help decision makers take informed decisions, prioritize investments, and develop effective energy policies that support economic excellence, environmental sustainability, and energy security [11]. While facing the challenges of energy transition and sustainability, it helps a nation hit a balance between its energy ambitions and more general economic growth goals.

Energy-Related Transactions Having Major Environmental Significance

Financial and operational tasks within the energy sector that have large environmental relevance—whether good or negative—are referred to as energy-related transactions of major environmental importance. These transactions may have a variety of negative effects on the environment, such as depletion of resources, disturbance of habitats, carbon emissions, and contamination of the air and water [12]. Encouraging environmental sustainability requires an understanding of and willingness to engage in these exchanges. Here are a few instances:

1. Fossil Fuel Exploration and Extraction: The exploration and extraction of fossil fuels, such as coal mining and oil and gas drilling, has a huge negative impact on the environment. In addition to contributing to climate change, they can result in habitat damage, water pollution, methane emissions, and other negative environmental effects.
2. Energy Infrastructure Development: The ecology may be seriously harmed by the construction of energy infrastructure, such as power plants, pipelines, and transmission lines. Large-scale projects may need the clearing of trees or the converting of land, harming ecosystems or affecting water supplies.
3. Renewable Energy Investments: On the positive side, transactions involving investments in renewable energy projects, including solar and wind farms, may have important environmental benefits. These transactions contribute to a reduction in greenhouse gas emissions, air and water pollution, and reliance on fossil fuels.
4. Energy Efficiency Initiatives: Transactions about energy efficiency initiatives and programs typically result in lower energy usage and greenhouse gas emissions. Examples include enhancing industrial procedures or putting energy-efficient technology in buildings.
5. Carbon Offset Transactions: By buying and selling carbon offsets, which represent emissions reductions or removals from initiatives like afforestation or renewable energy sources, the environmental effects of carbon emissions can be mitigated.
6. Emissions Trading: Similar to cap-and-trade systems, transactions in emissions trading markets have an impact on an industry's carbon footprint and may motivate efforts to cut pollution.

7. Environmental Impact Assessments (EIAs): Environmental impact studies are crucial for transactions involving energy-related developments. These assessments help in detecting potential environmental issues and informing decision-makers about the necessity of taking the proper mitigating measures.
8. Electric Vehicle (EV) Adoption: Transactions regarding the purchase and adoption of electric vehicles have an effect on the environment as compared to regular petrol or diesel-powered vehicles.
9. Biofuel Production and Trade: Depending on factors like changes in land use, water use, and the carbon intensity of production, transactions involving biofuels like ethanol and biodiesel can have complicated consequences on the environment.
10. Carbon Pricing Mechanisms: Transactions involving carbon pricing, like carbon taxes or emissions trading, have an impact on the financial incentives for reducing greenhouse gas emissions across a range of industries.
11. Natural Gas Imports/Exports: Natural gas import and export deals may have an impact on the world's energy markets and the environmental effects of using the fuel, including its potential to be a transition fuel with fewer emissions than coal.
12. Mining and Processing of Critical Minerals: Transactions involving the extraction and processing of vital minerals required for renewable energy technology (such as lithium for batteries) have an effect on the environment, harming habitats and using water.
13. Nuclear Energy Investments: Transactions in the nuclear energy sector could have a big impact on environmental issues, such as how radioactive waste is managed, and safety issues, such as the potential for accidents.

Making well educated decisions, encourage sustainable behaviors, and solving environmental problems like climate change, air and water pollution, and habitat preservation all depend on understanding and tracking the environmental importance of energy-related transactions. This knowledge can be used by individuals, businesses, and decision makers to create energy policies that support environmental sustainability objectives [13].

References

1. https://www.iea.org/reports/key-world-energy-statistics-2021.
2. https://www.iea.org/data-and-statistics.
3. https://unstats.un.org/unsd/energystats/.
4. https://www.ren21.net/wp-content/uploads/2019/05/GSR2022_Full_Report.pdf.
5. Dethlefs, T., Preisler, T., & Renz, W. (2015, November). A DER registry system as an infrastructural component for future smart grid applications. In International ETG Congress 2015; Die Energiewende-Blueprints for the new energy age (pp. 1–7). VDE.
6. https://data.oecd.org/energy.htm.
7. Renewables 2022 Global Status Report.
8. https://www.energyinst.org/statistical-review.

9. https://sdgstoday.org/.
10. https://www.stats.gov.sa/en/1042.
11. Energy Statistics Pocketbook 2023, United Nations Department of Economic and Social Affairs, March 2023, https://doi.org/10.18356/9789210012911.
12. Renewable Energy Data Book, 2018, US Department of Energy.https://www.nrel.gov/docs/fy20osti/75284.pdf.
13. International Recommendations for Energy Statistics, United Nations, 2018.

Importance of Energy Balance in Energy Statistics and Energy Policies

Yousef Al-Abdullah

Abstract From fueling our vehicles to powering our lights and HVAC systems, energy underpins our modern lives. With such paramount significance, the tracking of energy statistics is an often overlooked yet important activity. Entire agencies, including the United Nations Statistics Division and the International Energy Agency, have teams dedicated to tracking energy commodities and monitoring their global trade. This is done to inform audiences ranging from politicians to the general public. Additionally, energy statistics form the basis of many models, whether it be for the world or for a specific territory. Statisticians from these organizations and others gather energy statistics and present them in the form of energy balances. These balances provide a comprehensive view of how energy flows in a given territory, from the extraction of primary energy resources to their transformation into secondary energy products, and how energy is utilized in the territory's demand sectors, including industry, residential, commercial/services, agriculture, transport, and others. This chapter serves as a guide on how to develop energy balances.

Keywords Energy statistics · Energy flow · Oil · Fossil fuel · Power generation

Introduction

In our information-based economy today, statistics playing a more critical role than ever. This even more prescient today due to the highly interconnected nature of our world today, even though there are some trends to decoupling. These statistics are used from product traders to world leaders and, therefore, timeliness and accuracy are required to make any informed analysis. Due to correlation between energy and socio-economic conditions, timely and accurate energy statistics are even more crucial.

Y. Al-Abdullah (✉)
Energy and Building Research Center, Kuwait Institute for Scientific Research, Kuwait City, Kuwait
e-mail: yabdullah@kisr.edu.kw

K. J. Sreekanth (ed.), *Energy Efficiency Improvements with Emission Abatement for Energy Sustainability*, https://doi.org/10.1007/978-981-97-9480-5_2

Many institutions publish energy statistics, including the World Bank, United Nations Statistics Division (UNSD), and International Energy Agency (IEA). In particular, the UNSD and IEA publish energy balances that track the flow of *energy products* throughout a particular region or country. Energy products can be categorized as *fuels* as well as *energy*. Fuels can be defined as combustible fuels such coal, oil, or natural gas. Energy for statistical purposes strictly refers to heat and electricity. Another set of categories for energy products is whether it primary energy or secondary energy. *Primary energy products* include coal, crude, natural gas liquids (NGLs), and natural gas. They are unprocessed energy products that were extracted from reserves and mineral deposit. *Secondary energy products* include energy products that are derived from primary energy. For example, from the prime energy product of crude oil, it is possible to derive a multitude of oil derived products from motor gasoline to gas oil (diesel) to heavy fuel oil. Other secondary energy include heat, which is used in industrial processes and the heating of buildings, and electricity, which is foundation of our modern lives and powers our electronics as well as appliances, but also is used in modern industry. Additionally, energy products can be categorized as whether or not they are renewables, which include electricity generation from solar and wind, but also biofuels. The United States of America (USA) blends biofuels into motor gasoline to meet their energy needs for motor fuels. For an overview, Fig. 1 shows the terms energy statisticians readily use and how their categorization.

The tracking of all these energy products is necessary. It is required for statisticians to understand the processes, but also how energy flows throughout a territory. Within

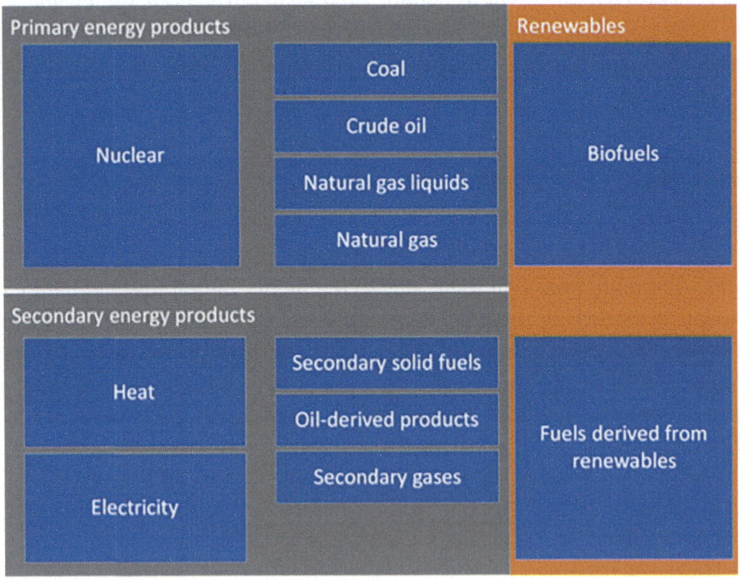

Fig. 1 Overview energy terms (inspired by IEA manual)

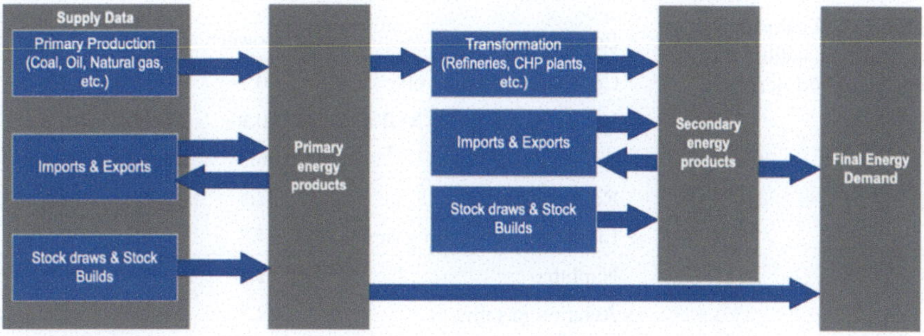

Fig. 2 Overview of energy flows in a given territory (inspired by IEA)

a territory, a nation will source primary energy products either from *primary energy production* (e.g., mining, oil extraction, etc.), *imports*, or withdrawals from current *stocks* of energy products in the country (also termed *stock draw*). Statisticians must also account for *exports* of primary energy products to other territories as well as any primary energy that goes into stocks, otherwise known as *stock builds*.

Primary energy products can be utilized to fulfill demand directly (*final energy demand*), but they are also processed into secondary energy products. For example, crude oil is transformed into a variety of secondary oil-derived products (motor gasoline, gas oil, heavy fuel, etc.) at oil refineries. Another transformation process is combined heat and power (CHP) or electricity plants, where even secondary fuels such as heavy fuel oil are burned to produce electricity and heat for consumers. Thus, energy products could undergo multiple transformation processes before meeting final energy demand. Furthermore, secondary energy products could also be imported or exported as well as stored or withdrawn from a territory's stocks. This general flow of energy products can be seen in Fig. 2.

All of these energy products and flows are tracked via energy balances. The UNSD, [1], and IEA, [2], provide guidelines on energy balances and this chapter the focus shall be on energy statistics for crude oil and oil derived products, which are the most widely trade products in the world, as well as natural gas.

Construction of Energy Balances

Tracking of energy products in begins with statisticians acquiring data from energy players from within the territory that is being studied. For construction of energy balances, two main metrics for energy statistics for each product need to be collected: (1) the amount (usually in physical units) an energy product that goes through each flow and (2) the net calorific values for each energy product. From these two data types, a statistician can then process the data into energy balance such that all energy products and flows are reported in a single common energy unit.

Table 1 Net calorific values for crude oil, natural gas liquids, and oil-derived products

Energy product	Lower*	Default*	Upper*
Conventional crude oil	40.1	42.3	44.8
Natural gas liquids (NGLs)	40.9	44.2	46.9
Refinery feedstocks	36.3	43.0	46.4
Refinery gas	47.5	49.5	50.6
Liquefied petroleum gases (LPG)	44.8	47.3	52.2
Naphtha	41.8	44.5	46.5
Aviation gasoline	42.5	44.3	44.8
Motor gasoline	42.5	44.3	44.8
Gasoline-type jet fuel	42.5	44.3	44.8
Kerosene-type jet fuel	42.0	44.1	45.0
Other kerosene	42.4	43.8	45.2
Gas oil/diesel	41.4	43.0	43.3
Fuel oil	39.8	40.4	41.7
White spirit	33.7	40.2	48.2
Lubricants	33.5	40.2	42.3
Paraffin waxes	33.7	40.2	48.2
Petroleum coke	29.7	32.5	41.9
Bitumen	33.5	40.2	41.2
Other oil products	33.7	40.2	48.2

* All units in terajoules (TJ) per thousand metric tons

The amount of energy a particular energy product. Original units obtained from data sources could come in a variety of unit types. For crude oil and oil-derived products, it is not unusual that the data provided from various energy players in a given territory to be provided in volumetric units. However, this could be there could be slight issues when converting the physical units into energy units because in particular for oil, lighter products are produced at refineries and therefore a mismatch could occur. Hence, physical units in mass are more desirable.

Calorific values the other data type needed for building an energy balance is also critical to collect from data providers in the energy industry. Calorific values allow statisticians to convert physical quantities into energy units. It is best practice to collect from the data source. However, at times this cannot be done. Therefore, the IEA and UNSD provide calorific values. In practice, net calorific values (NCV) should be utilized over gross calorific values (GCV) due to the NCV being the amount of heat that is available for combustion, while the gross includes latent heat that is not captured. For reference, NCV and GCV are referred to as low heat value and high heat value. In Table 1, the NCV from [1] is shown. Note that in the Table 1, a range is provided.

Acquiring the data begins with seeking out local sources within the energy industry of that particular territory. However, there are international sources that can be more easily accessed for start the construction of an energy balance. For example, Joint Oil Database Initiative (JODI) receives data from local data sources and provides data for free on their website [3]. Furthermore, the IEA utilizes this platform as part of their work in constructing their energy balances. Utilizing JODI, an energy statistician can pull data on any country that provides data and is provided monthly. A simplification of the data needed is in Table 2. As an example, Kuwait is being used.

Utilizing the calorific values from Table 1 and the reported totals in Table 2, we will transform each column (flow) the data into a common energy unit, million tons of oil equivalent (Mtoe), utilizing Eq. (1):

$$\begin{aligned} \textit{Energy flow} &= \textit{Physical Units (thousands of tons)} \\ &\times \textit{Calorific Value}\left(\frac{TJ}{\textit{thousand tons}}\right) \\ &\times \textit{Conversion to Mtoe}\left(\frac{1\,Mtoe}{41868\,TJ}\right) \end{aligned} \tag{1}$$

The result of transforming the data can be viewed in Table 3.

The same process can be utilized for secondary oil products. For simplicity, a selection of the products will be utilized and the initial data total for the year 2021 along with the units transformed into Mtoe are provided in Table 4.

Table 2 Data download from JODI database for Kuwait

Month	Production	Imports	Exports	Direct use	Stock change*	Refinery intake
(All units in thousands of metric tons)						
Jan-21	9,941.38	0.00	7,290.34	128.28	-46.76	2,569.79
Feb-21	8,994.76	0.00	6,816.55	77.24	−177.66	2,278.62
Mar-21	9,949.93	0.00	7,504.14	42.76	−180.00	2,582.62
Apr-21	9,628.97	0.00	7,009.66	82.76	52.55	2,482.76
May-21	10,069.66	0.00	7,333.10	106.90	64.14	2,565.52
Jun-21	9,864.83	0.00	7,013.79	132.41	235.86	2,482.76
Jul-21	10,360.41	0.00	7,593.93	158.21	42.76	2,565.52
Aug-21	10,454.48	0.00	7,380.14	171.03	140.83	2,762.21
Sep-21	10,237.24	0.00	7,328.28	124.14	−127.03	2,913.10
Oct-21	10,702.48	0.00	7,384.41	0.00	336.69	2,980.28
Nov-21	10,477.24	0.00	7,386.21	0.00	−12.00	3,103.45
Dec-21	10,899.17	0.00	7,752.14	0.00	−102.48	3,249.66
Total	121,580.55	0.00	87,792.69	1,023.72	226.90	32,536.28

* From JODI a positive number indicates a stock build and a negative number indicates a stock draw

Table 3 Converted JODI data of crude oil in Mtoe

Units in Mtoe	Production	Imports	Exports	Direct use	Stock changes	Refinery intake
Total	122.84	–	88.70	1.03	0.22	32.87

Table 4 Secondary oil product data from JODI and the resulting conversion to Mtoe

Fuel	Units	Refinery output	Imports	Exports	Stock changes	Demand
Naphtha	Thousands of metric tons	5,701.60	0	5,216.25	–2,412.36	2,901.83
Motor and aviation gasoline		2,611.53	194.71	0	–419.76	3,228.47
Kerosene		6,861.90	0	6,242.79	68.61	551.77
Gas/diesel oil		9,374.59	0	7,624.46	–996.89	2,755.14
Fuel oil		5,092.00	1,627.85	851.85	90.92	5,784.31
Other oil		495.60	16.60	57.77	–67.35	523.21
Fuel	Units	Refinery output	Imports	Exports	Stock changes	Demand
Naphtha	Mtoe	6.06	0	5.54	–2.56	3.08
Motor and aviation gasoline		2.76	0.21	0	–0.44	3.42
Kerosene		7.18	0	6.53	0.07	0.58
Gas/diesel oil		9.63	0	7.83	–1.02	2.83
Fuel oil		4.91	1.57	0.82	0.09	5.58
Other oil		0.48	0.97	0.02	0.06	–0.06

Pairing the crude oil with the secondary oil products, a portion of the energy balance can be built for Kuwait. Filling the energy balance table, requires careful attention. Beginning with crude oil conversions from Table 3, matching the flows is required in Table 5.

First, “production”, “imports”, “exports”, and “stock changes” from Table 3 are listed in Table 5 as “primary production”, “imports”, “exports”, “stock changes”. However, notice that stock changes that the sign changes for the energy balance and is due to the need to indicate that this amount of crude oil was withdrawn from the territory and stored for that given year. The next like on “total energy supply” quantifies the amount of energy total or for a particular fuel that stayed within a territory. Subsequently, the line on “statistical difference” indicates any issues with the data and compares total energy supply with other flows below. These “statistical differences” are calculated via Eq. (2).

Table 5 Construction of an oil balance from JODI data—all units in Mtoe

Kuwait 2021 units Mtoe	Crude oil	…	Naphtha	Motor gasoline	Kerosene	Gas oil	Fuel oil	Other oil	…	Total
Primary production	122.84									122.84
Imports				0.21			1.57	0.02		1.79
Exports	–88.70		-5.54	0.00	-6.53	-7.83	-0.82	-0.06		-109.48
International aviation bunkers										
International marine bunkers										…
Stock changes*	–0.22		2.56	0.44	–0.07	1.02	–0.09	0.06		3.72
Total primary Energy supply	33.92	…	–2.98	0.65	–6.60	–6.81	0.66	0.03		18.87
Statistical difference	0.02					–0.01	-0.01			0
Transfers										
Transformation	–33.90	…	6.06	2.76	7.18	9.63	4.91	0.48		–2.88
…										
Oil refineries	–32.87		6.06	2.76	7.18	9.63	4.91	0.48		–1.85
Energy industry own use	–1.03									
Losses										
Total final energy consumption	0		3.08	3.42	0.58	2.83	5.58	0.50		15.99

$$\begin{aligned} \textit{Statistical Differences} &= \textit{Total Energy Supply} \\ &\quad + \textit{Transfers} + \textit{Transformation} \\ &\quad - \textit{Total final energy consumption} \end{aligned} \tag{2}$$

Proceeding, "transfers" are used to indicate when certain fuels are reclassified as other fuels. Potentially, light naphtha could be transferred and used as motor gasoline. The next heading in the energy balance table (Table 5) is total amount a particular fuel is used or is produced by transformation processes. In this example, crude oil is the input for oil refineries. Notice, that the input is negative, and this is the total amount of crude oil that was used in the refinery to produce the secondary oil products (i.e., naphtha, motor gasoline, kerosene, gas oil, fuel oil, and other oil). Notice that the for the secondary products are all positive because they are produced by oil refining transformation process. Continuing to track crude oil, "energy industry own use" of crude oil is taken from the JODI data in Table 3 from the "direct use" value.

The final line for "total final energy consumption" indicates the total consumption of each energy product in the territory. Notice that crude oil has zero final energy consumption, and this is the case because oil is refined in required fuels for consumers. For example, motor gasoline for transportation. It is desirable to track energy demand in the various sectors of the economy of the territory, including industry, transport, agriculture, fishing, forestry, services, residential, and non-energy use.

Furthermore, for tracking industrial demand, a statistician would desire to collect data on the following major industries: (1) iron and steel, (2) chemical and petrochemical, (3) non-ferrous metals, (4), non-metallic minerals, (5) transport equipment, (6) machinery, (7) mining and quarrying, (8) food and tobacco, (9) paper, pulp and printing, (10) wood and wood products, (11) textile and leather, as well as (12) construction. Sectoral based energy consumption is not tracked by JODI. Hence, these energy flows were not specified in Table 5. However, to see an example, it is recommended to visit [4], which reports energy balances for all countries. Additionally, some assumptions throughout this process and statistician will have to make assumptions based the knowledge of the territory being study. For example, JODI reports production of liquefied petroleum gases (LPG) from refineries in Kuwait. However, only a portion of LPG is produced at refineries and is produced at NGL and gas processing facilities in the country. Since NGL production was not listed for Kuwait on JODI, it was not included as part of Table 5. Further data collection would be needed to complete the energy balance table above, especially tracking sectoral based energy demand.

Tracking of Natural Gas

Utilizing the JODI data once again [3], a statistician can begin including natural gas in the energy balances. The natural gas data from JODI includes several energy flows and for the example presented it is possible to attain production, imports of LNG), exports, gross inland deliveries, and deliveries to electricity plants in Kuwait. Note that JODI reports deliveries to electricity and heat plants, however, Kuwait does not have any heat plants. Kuwait does utilize thermal processes for desalination, but this is not accounted for by JODI. A summary of natural gas statistics from JODI in Table 6 [3].

Processing the data requires to convert natural gas into the same energy unit as chosen before, Mtoe. The conversion is similar to Eq. (1). However, the units that are reported are in million standard cubic meters. Therefore, one must either convert cubic meters to metric tons or in this case it is possible to utilize the net calorific value of 39.02 TJ per million cubic meters from [1]. The results are in Mtoe are presented in Table 7.

Adding the data in Mtoe for natural gas is similar to that of oil detailed above. In Table 8, the energy balance with natural gas is added. As one can see. Production of natural gas is filled into the energy flow row for primary production. This is followed by imports of LNG in the row labeled imports. These two values sum to

Table 6 JODI data summarized

Units in million standard cubic meters	Production	Imports (LNG)	Exports	Gross inland deliveries (Observed)	Of which: electricity and heat generation
Jan2021	1,262	0	0	982	289
Feb2021	1,163	70	0	946	292
Mar2021	1,344	422	0	1,039	331
Apr2021	1,310	617	0	963	264
May2021	1,367	670	0	1,026	262
Jun2021	1,363	673	0	1,048	346
Jul2021	1,381	802	0	1,140	404
Aug2021	1,375	1,053	0	1,164	368
Sep2021	1,311	964	0	1,089	305
Oct2021	1,297	830	0	1,098	304
Nov2021	1,355	481	0	1,089	301
Dec2021	1,406	441	0	1,145	321
Total	15,934	7,023	0	12,729	3,787

Table 7 Natural gas converted into Mtoe

Units in Mtoe	Production	Imports (LNG)	Exports	Gross inland deliveries (Observed)	Of which: electricity and heat generation
Total	14.85	6.55	–	11.86	3.53

the total primary energy supply. Next, one must account for gross inland deliveries and deliveries to electricity plants. Therefore, first a new flow is added for electricity plants, which for 2021 Kuwait utilized 3.53 Mtoe for electric power generation according to JODI. The remainder of gross inland deliveries is not accounted for in JODI and according to its definition, some of these deliveries could go to the energy industry or it could go to final demand. This data needs to be attained from local data sources. For simplicity at this point, the remainder is accounted for in total final energy consumption for now. It would be desirable to also attain the electricity generation from natural gas generation in the territory and from, [5], it is stated that 43.8 Terawatt-hour (TWh) was generation from natural gas. This is equivalent to 3.77 Mtoe and this filled into the for electricity. There are more statistics to be gathered regarding electricity in this territory. However, for this example is to track natural gas. Notice that another issue arises. The amount of electricity produced is greater than the fuel input and this is not correct. The efficiency of natural gas generation has a range topping out at approximately 70% efficiency. This issue arises at times, especially when comparing disparate international sources. Under these circumstances, it is necessary to attain local data to obtain the correct amount of natural gas for power generation and the electricity generated fueled exclusively by natural gas. In the past, Kuwait's Ministry of Electricity, Water, and Renewable Energy published statistics on fuel consumption and these numbers could be attained. However, recently these reports do not include fuel consumption at power stations [6].

Metrics

Energy balances allow statisticians to see energy flows in the studied territory. There are some main energy indices that can be attained from the energy balance. One such example is total primary energy supply (TPES), which accounts for all primary energy products that are required to fulfill local energy demand. As shown above, this metric includes all primary energy production, imports, exports, deliveries to international aviation and marine bunkers, and stock changes. Similarly, energy balances could report total energy supply (TES), which includes the production of secondary energy products. It is important to distinguish between the two depending upon data source. Additionally, total final energy consumption highlights the energy that was allocated to consumers. The energy industry and transformation sectors of a territory have their own consumption and it is necessary to distinguish. Finally, statistical differences quantify the difference supply and utilization of an energy product. When there are large statistical differences in the energy balance tables, it is necessary to attain local data. This does occur at times, especially when comparing data from international sources.

Summary

Energy balances are useful tools for wide range of audience and provide users with the ability to track the amount of energy used in a given territory and how the energy was utilized within the territory. These statistical reports are used by commodity traders to policy makers. However, their formulation requires timely and accurate data. As seen in the example presented not all can be fully verified and, therefore, local data from willing sources must be attained and processed.

Table 8 Energy balance with natural gas

Kuwait 2021 units Mtoe	Crude oil	Natural gas	…	Naphtha	Motor gasoline	Kerosene	Gas oil	Fuel oil	Other oil	…	Electricity	Total
Primary production	122.84	14.85										137.69
Imports		6.55			0.21			1.57	0.02			8.34
Exports	–88.70			–5.54	0.00	–6.53	–7.83	–0.82	–0.06			–109.48
International marine bunkers												0.00
International aviation bunkers												0.00
Stock changes	–0.22			2.56	0.44	–0.07	1.02	–0.09	0.06			3.72
Total primary Energy supply	33.92	21.40		–2.98	0.65	–6.60	–6.81	0.66	0.03			40.26
Statistical difference	0.02	9.53		0.00	0.00	0.00	-0.01	–0.01	0.00			9.53
Transfers												
Transformation	–33.90	–3.53		6.06	2.76	7.18	9.63	4.91	0.48		3.77	–2.65
…												
Electricity plants		–3.53									3.77	0.24
Oil refineries	–32.87			6.06	2.76	7.18	9.63	4.91	0.48			–1.85
Energy industry own use	–1.03											
Losses												
Total final energy consumption		8.33		3.08	3.42	0.58	2.83	5.58	0.50		3.77	24.32

References

1. *International recommendations for energy statistics (IRES).* New York: United Nations, 2018.
2. International Energy Agency, *Energy Statistics Manual.* OECD, 2004. https://doi.org/10.1787/9789264033986-en.
3. "Beyond 20/20 WDS—Reports." Accessed: Oct. 10, 2023. [Online]. Available: http://www.jodidb.org/ReportFolders/reportFolders.aspx?sCS_referer=&sCS_ChosenLang=en.
4. "UNSD—Energy Statistics." Accessed: Oct. 08, 2023. [Online]. Available: https://unstats.un.org/unsd/energystats/pubs/balance/.
5. "International—U.S. Energy Information Administration (EIA)." Accessed: Oct. 11, 2023. [Online]. Available: https://www.eia.gov/international/analysis/country/KWT.
6. "Electricity book 2022.pdf."

Energy Efficiency and Emission Abatement—an Indian Perspective

V. R. Rajesh

Abstract India is having third position in the emission of greenhouse gases across the globe following China and the United States. Thermal power plants based on coal, various industries, transportation, paddy fields and domestic animal farms are the contributing sources of carbon emission in the country. However, India's greenhouse gas emissions per head are considerably lower than the global average, at just 2.8 tCO_2e. In contrast, per capita emissions in China and the United States were 9.6 and 17.6 tCO_2e, respectively. Furthermore, India's contribution to historical cumulative CO_2 emissions is also much lower than other major economies, at less than four percent. Also, India is a leader in energy efficiency measures, thanks to the innovative practices of the Bureau of Energy Efficiency (BEE). The BEE has implemented a number of initiatives to promote energy efficiency, including the Standards and Labeling (S&L) program, the Energy Efficiency Building Code (EEBC), the Energy Efficiency Labeling Program (EELP), and the Energy Efficiency Financing Program (EEFP). India has also taken steps to reduce its carbon dioxide emissions through the Perform, Achieve and Trade (PAT) scheme. This scheme allows companies to trade emission reductions, which provides an incentive for companies to invest in energy efficiency measures. The PAT scheme has been implemented from the year 2017 on an annual basis in rolling cycle pattern. Most energy intensive sectors such as Cement, Iron and Steel, Fertilizer, Thermal Power Plants, Refineries, Petrochemicals, Railways and others are covered under this scheme till the year 2020. 17 Mtoe (Million Tonnes of Oil Equivalent) energy savings have been noticed through this initiative by the Govt. of India and of about 87 million tonnes of CO_2 emission per year have been resulted in this way. By the usage and implementation of energy efficiency measures, 15% of extra energy demand is eliminated and resulted in the reduction of 300 $MtCO_2$ emissions in the period of 2000–2018. The country has pledged in the Paris agreement a 33–35% reduction in emissions by 2030 compared to 2005 levels. India's energy efficiency efforts are a model for other countries around the world. India's success shows that it is possible to reduce energy consumption and greenhouse gas emissions while also improving economic growth. At the COP26

V. R. Rajesh (✉)
Department of Technical Education, Government Polytechnic College Palakkad, Palakkad, Kerala, India
e-mail: rajeshvr@gptcpalakkad.ac.in

K. J. Sreekanth (ed.), *Energy Efficiency Improvements with Emission Abatement for Energy Sustainability*, https://doi.org/10.1007/978-981-97-9480-5_3

international climate change conference, India announced more aggressive targets, including the goal of achieving net-zero emissions by 2070.

Keywords Greenhouse gas · Emission · Electrical energy · Carbon dioxide · Climate change

Introduction

Energy is an inevitable component over every phase of day-to-day affairs across different sectors of activities. A country's economic growth is highly dependent on the amount of energy expended. Reducing the wastage of energy for executing a task is termed as energy efficiency. Or, in other words, performing the same task with a smaller amount of energy is said to be more energy efficient. Efficient use of energy is the highest and cost-effective way to keep the capital, trim down greenhouse gas emission, generate jobs, and meet growing energy demand. Efficient usage of energy substantially cut down our dependence on fossil fuels, which are a major source of greenhouse gas emissions. We can also save money on our energy bills, which can free up money for other things. In addition, energy efficiency can create jobs in the manufacturing and installation of energy-efficient products and services. Finally, energy efficiency can help us meet our growing energy demand without building new power plants, which can save money and reduce environmental impacts.

Here are some examples of how we can use energy more efficiently at root level:

- Use energy-efficient appliances and light bulbs.
- Weatherize your home or business.
- Drive a fuel-efficient car.
- Use public transportation or carpool.
- Conserve water.
- Recycle and compost.
- Support renewable energy sources.

By using energy more efficiently, we can all help to protect the environment and create a more sustainable future. The reduction of energy usage ultimately results in reduced carbon emissions and thereby protecting the environment in a broad sense.

Energy efficiency is a key strategy in combating climate change, which is now more pressing than ever due to the recent increase in emissions and the limited time we have to achieve mitigation goals. Energy efficiency is one of the most important ways the world can meet its energy needs while using less energy, which is essential for most of the IPCC's (Intergovernmental Panel on Climate Change) GHG (Greenhouse Gases) emissions pathways that limit global warming to 1.5°C [1].

The Bureau of Energy Efficiency (BEE) was established through Energy Conservation Act of India (2001). BEE was assigned the job of cutting down the energy intensity of the economy. Relatively latest technologies namely E-mobility, Fuel

Cell Vehicles, integration of renewable energy conversion & storage, net zero buildings, district cooling, smart meters, internet of things, active appliance feedback, block-chain technologies etc. were listed under the BEE's National Strategic Plan for the reduction of carbon footprint of the economy and thereby increase the energy efficiency.

National Mission on Enhanced Energy Efficiency (NMEEE) is the highlight of India's National Action Plan on Climate Change (NAPCC) among the eight missions. The major objective of NMEEE is said to be the reduction of energy intensity with respect to economy similar to BEE. Four initiatives have been put forward by NMEEE to improve energy efficiency in the country. They are as follows:

1. Perform Achieve and Trade (PAT)
2. Market Transformation for Energy Efficiency (MTEE)
3. Energy Efficiency Financing Platform (EEFP)
4. Framework for Energy Efficient Economic Development (FEEED).

In order to reduce the carbon dioxide emissions, the Indian government takes measures to save energy by various means through the Perform, Achieve and Trade (PAT) scheme. However, like any other investments, carbon abatement costs are also more upfront, but are expected to give positive returns in the long run only. With the implementation of various strategies and measures, the country is trying to target a decrease of about 33–35% in the carbon emissions of its economy within the period of 2005–2030. Renewable energy resources utilization in the generation of electricity is also targeted to be about 40% by the year 2030. Through the reduction of deforestation it is planned to create an additional cumulative carbon sink of 2,500–3,000$MtCO_2e$ (Million tonnes of Carbon Dioxide equivalent) by the year 2030. The first PAT cycle was implemented between 2012 and 2015, which led to a savings of 31$MtCO_2e$, roughly 1% of the total annual emissions in the country.

The performance of buildings with regard to energy consumption during its construction and operation and environmentally friendly manufacturing are continuously evaluated by the nation's smart grid mission. India's emissions reduction rate increased to 3% annually between 2016 and 2019, up from 1.5% between 2014 and 2016. This was the fastest reduction to date, and was largely due to the government's push towards renewable energy, even as fossil fuels continue to dominate the energy mix.

Greenhouse Gases and Sources

In terms of energy consumption, India ranks third globally, accounting for 569 Mtoe. The industrial sector consumes a significant portion of this, approximately 318 Mtoe, representing 56%. Notably, carbon dioxide (CO2) is a greenhouse gas known to trap heat in the atmosphere, contributing to global warming. The greenhouse effect works when solar radiation hits the Earth's surface, some of the heat escapes into space while the rest is trapped by CO_2 and other greenhouse gases. This trapped heat warms the planet, causing global warming. The ramifications of global warming are far-reaching, affecting all natural ecosystems and by association, every industry and individual across the globe. The ongoing climate crisis has resulted in a myriad of environmental and health issues, such as rising sea levels, more frequent and severe weather events, and alterations in plant and animal life. In India, the annual mean temperature has witnessed a significant increase of approximately one degree Celsius since 1850, reflecting the urgent need for action to mitigate the impacts of climate change.

The foremost cause of greenhouse gas emissions in India is the combustion of coal, used directly and indirectly, primarily for electricity generation and transportation [2]. Coal is India's primary energy source, accounting for over 70% of the country's power mix. In 2021, coal combustion in India produced 1.8 $GtCO_2$ (Giga tonnes of Carbon Dioxide). In India, the electric power sector is the biggest contributor to greenhouse gas (GHG) emissions, primarily due to the massive amounts of carbon dioxide (CO_2) emitted by coal-fired power plants. The agriculture sector is the second-largest contributor, releasing significant quantities of methane (CH_4) from rice paddies and livestock. CH_4 is a potent GHG, with its 100-year global warming potential roughly 30 times greater than that of CO_2.

India's reliance on coal is a major obstacle to the country's climate goals. The government has implemented various measures to curb its emissions intensity to the levels of 33–35% in 2030 compared to 2005, but this will be difficult to achieve without a significant shift away from coal. The government has taken some steps to reduce its reliance on coal, such as promoting renewable energy and energy efficiency. However, these measures have not been enough to offset the growth in coal consumption.

India will need to take more ambitious steps to reduce its coal consumption if it is to meet its climate goals. This could include investing in new technologies such as carbon capture and storage, or phasing out coal-fired power plants. The government will also need to address the social and economic impacts of a transition away from coal. Coal mining and power generation provide employment for millions of people in India. A transition away from coal will need to be managed carefully to ensure that these workers are not left behind.

In 2015, India's greenhouse gas (GHG) emissions, measured in carbon dioxide equivalents ($MtCO_2e$), reached 3,571 million tonnes, as per data compiled by the Potsdam Institute for Climate Impact Research (PIK). This represented a more than threefold increase since 1970. However, India's per capita emissions stood at 2.7tCO_2e in 2015, which was approximately one-seventh of the US figure and less than half of the global average of 7.0tCO_2e.

Historically, agriculture was the predominant contributor to greenhouse gas (GHG) emissions in India. However, while emissions from this sector have experienced a 30% increase since 1990, emissions originating from the electricity and heat sector have exhibited a significantly higher growth rate, increasing five-fold. Evidently, electricity generation and thermal energy usage constitute the primary sources of GHG emissions by sector in India, as indicated by the available data.

Preventive Measures

Within the Indian context, a multifaceted policy and institutional framework has been established to facilitate the promotion of energy efficiency. The Ministry of Power (MoP) assumes the primary responsibility of formulating policies for the electricity sector. Functioning as a subordinate entity of MoP, the Bureau of Energy Efficiency (BEE) is entrusted with the implementation and regulation of energy efficiency policies and programs, as outlined in the ambit of the Energy Conservation Act, 2001. Subsequently amended in 2010, this act serves as the bedrock of India's energy efficiency policy framework. Moreover, the National Mission on Energy Efficiency, one of eight missions encompassed within the 2008 National Action Plan on Climate Change, stands as a testament to India's steadfast commitment to energy efficiency.

The roles of MoP and BEE are critical in the collection, compilation, and subsequent analysis of energy efficiency data. Furthermore, the development of indicators is undertaken to monitor the effectiveness of implemented policies and programs and to track progress over time. Currently, India is actively engaged in efforts to disaggregate energy consumption by subsector and end use, with the aim of gaining a more granular understanding of energy consumption patterns.

The Energy Conservation Act mandates that states establish dedicated entities to facilitate the implementation of energy-saving plans and the creation of State Energy Conservation Funds (SECFs). Presently, twenty-eight states maintain SECFs, and the majority have allocated funding to them. States play a pivotal role in establishing and enforcing comprehensive energy-saving regulations, developing the necessary skills and knowledge among the populace, formulating plans, initiating and executing projects, providing financial assistance to eligible entities, and fostering markets for energy-efficient products and services.

In efforts to reduce CO_2 emissions from industries and establishments, the Ministry of Power has implemented various measures. One such key program is the Perform, Achieve, and Trade (PAT) Scheme. It targets large industries and establishments, aiming to make energy savings more cost-effective through technology upgrades or internal initiatives. The scheme establishes mandatory targets for major energy consumers, and any surplus energy saved is issued as tradable Energy Saving Certificates. Based on their energy usage and potential for savings, various industries and establishments are assigned individual energy efficiency targets.

By the year 2020, the ambit of the scheme had been formally expanded to encompass thirteen energy-intensive sectors in the nation, including Cement, Iron, and Steel, Fertilizer, Thermal Power Plants, Refineries, Petrochemicals, Railways, and others. This endeavor is currently yielding energy savings of approximately 17 Mtoe (Million Tonnes of Oil Equivalent) and has resulted in the mitigation of approximately 87 million tons of carbon dioxide annually, a figure that closely approximates the total carbon dioxide emissions of nations such as Bangladesh.

Consumer awareness on the usage of energy efficient equipment and appliances and energy saving potential is enabled by BEE through Standards and Labeling (S&L)

program. The rising demand for electrical energy could be optimized by switching to energy efficient devices in household systems.

In the electricity sector, there are two main approaches to reducing emissions: one is to improve efficiency by using less energy and materials, and the other is to switch to renewable energy sources. Improving efficiency can be achieved through a variety of means, such as using more energy-efficient appliances and lighting, improving insulation in buildings, and using more efficient transportation. Switching to renewable energy sources, such as solar and wind power, can also help to reduce emissions. Both of these approaches are important for reducing emissions from the electricity sector. However, improving efficiency is often the more cost-effective option in the short term. Switching to renewable energy sources can be more expensive, but it can also provide long-term benefits, such as reducing reliance on fossil fuels and improving energy security.

The food industry and agriculture can reduce CO_2 emissions by lowering meat and dairy consumption, particularly beef. Reducing meat and dairy intake can also help to reduce methane emissions. While chicken or fish still produce carbon emissions, they produce less than beef. Regenerative agriculture, which involves using regenerative soil practices, can significantly improve the environment and is a great way to combat global warming. It is an inexpensive and organic way to improve the atmosphere by sequestering large amounts of CO_2 emissions. Sustainable intensification is a strategic approach to crop production that increases yields without harming the environment. It does this by using less land, preventing deforestation, and improving ecosystems.

Food waste accounts for around six percent of global greenhouse gas emissions. While much of this waste comes from consumers, over half of it is lost in supply chains. Reducing spoiled food in processing and waste in restaurants and hotels could help create a more sustainable future. Transparency in the supply chain, especially for food, has become increasingly important in recent years, not only for health reasons but also for best environmental and labor practices. Regenerative agriculture, transparency, and not discarding fruits and vegetables that are not aesthetically pleasing will all be beneficial in this endeavor.

Electric vehicles and liquid natural gas (LNG) vehicles emit less carbon dioxide than traditional petroleum gasoline and diesel vehicles. The plan is to decarbonize new transportation by not selling petroleum-burning vehicles anymore. This will have to be done soon in order to reach net zero emissions by 2050.

Abatement of Emissions Through Energy Efficiency

The reduction of greenhouse gas (GHG) emissions in order to lower the amount of GHGs polluting the atmosphere is generally termed to be emissions abatement. Burning of fossil fuels is the main culprit behind carbon dioxide emission into the atmosphere along with other natural sources such as biological decay. The risks associated with the GHGs in the atmosphere are a lot but not limited to: loss of nutrients from the crops (soil pollution), dying ecosystems and extinct animal species, respiratory diseases due to smog (air pollution), flood resulting from excessive rainfall which in turn is caused by overheating of atmosphere, heat waves and melting of polar ice caps.

Energy conservation and energy efficiency are closely related but distinct concepts. Energy conservation is the reduction of energy consumption through more efficient energy use. Energy efficiency is the reduction of energy losses by using advanced technologies. Energy conservation can be achieved by using less energy-intensive appliances, turning off lights when they are not in use, and weatherizing homes and businesses. Energy efficiency can be achieved by using energy-efficient appliances, improving insulation, and using renewable energy sources. Both energy conservation and energy efficiency are important for reducing our reliance on fossil fuels and protecting the environment. Material efficiency and energy efficiency can reinforce each other. For example, using less steel in an automobile saves energy in production and improves fuel efficiency. Reducing demand for energy-intensive materials and recycling lowers energy demand. Recycling metals uses 60–90% less energy than primary production.

The Republic of India maintains a comprehensive and diverse portfolio of policies and programs related to energy efficiency. In the recent past, the nation has embarked upon the initiation of innovative and impactful programs in this domain. Furthermore, a significant number of energy efficiency projects have benefited from the support of bilateral and multilateral programs. The draft National Energy Policy formulated by NITI Aayog emphasizes the pivotal role of energy efficiency in achieving four primary objectives: ensuring energy access at affordable prices, enhancing energy security and independence, promoting greater sustainability, and facilitating economic growth.

The significance of energy efficiency for India's goal of reducing emission intensity cannot be overstated. The Indian government's efforts in recent years to expand access to electricity and clean cooking have been exemplary. Moving forward, the government has ambitious plans to provide its citizens with reliable, reasonably priced, and environmentally friendly energy. Due to the crucial role of energy in economic growth, it is imperative to implement cutting-edge and innovative policies to reduce wasteful energy consumption across all sectors. The Bureau of Energy Efficiency, guided by this understanding, has developed the National Strategy Plan known as Unlocking National Energy Efficiency Potential (UNNATEE). The study concludes that India has the potential to save 86.9 Mtoe under the moderate implementation of EE (Energy Efficiency) programs and 129 Mtoe under the ambitious implementation of EE programs by 2031, which represents a 15% reduction in energy demand compared to the conventional approach to energy conservation.

The aggregate values of energy savings attained for all these programs over the last three years are juxtaposed with energy savings objectives for diverse demand sectors and depicted in Table 1 [3].

From Table 1 it is evident that domestic and industrial sectors have performed well in achieving the energy savings targets in the initial years of implementation even though other sectors like transportation and agriculture are running far behind. This

Table 1 Energy savings potential and achieved targets

Sector	Energy savings potential (Mtoe) till 2031 from 2016	Pro-rata energy savings targets (Mtoe) for 2017–2020	Total savings (Mtoe) for 2017–2020
Agriculture	5.7	1.22	0.75
Commercial	4.9	1.05	0.15
Domestic	12.1	2.6	7.73
Industrial	47.5	10.19	14.98
Transport	15.8	3.39	1.44
Total	86.8	18.45	25.05

shortfall, anyhow, is not a thought of concern since the technological advancements in those areas are being implemented in a slow pace when compared to industrial and domestic sectors where energy consumption is quite a large. It is expected that all the sectors listed above are capable of achieving the targets within the period.

The Perform, Achieve, and Trade (PAT) program constitutes a market-based mechanism intended to augment the cost-effectiveness of enhancements in energy efficiency within energy-intensive industries and facilities of substantial size. This is accomplished through the certification of quantifiable energy savings that can be traded. Within this mechanism, the government establishes specific targets for industries to reduce their Specific Energy Consumption (SEC). Entities that successfully achieve and surpass these targets receive Energy Saving Certificates (ESCerts). Conversely, those industries falling short of the established goals either incur financial penalties or acquire ESCerts from entities that have exceeded their targets and accumulated surplus ESCerts.

The implementation of policies aimed at conserving energy and promoting energy efficiency encompasses various options, including enhancing the energy efficiency of appliances through the Standards & Labeling (S&L) program. The fundamental objective of the S&L program is to ensure the removal of highly inefficient products from the market and to provide consumers with standardized labels that facilitate informed decision-making, encouraging the acquisition of increasingly more efficient appliances. Commencing in 2006, the S&L program initially introduced voluntary labeling for refrigerators and fluorescent tube lights. Subsequently, in 2009, mandatory labeling was implemented for a diverse range of appliances, which included room air conditioners, tubular fluorescent lamps, frost-free refrigerators, and distribution transformers. At present, the program encompasses an extensive array of 26 appliances, with 10 subject to mandatory labeling and 16 under voluntary labeling.

Established standards delineate energy consumption limitations for manufactured products, encompassing minimum proficiency thresholds or categorical prohibitions on subpar products. Additionally, these standards may incorporate standardized testing protocols to accurately quantify energy performance. Energy efficiency labels serve as informative documentation, affixed to products, to convey their energy performance characteristics, typically in terms of energy consumption, efficiency, or cost. Such labels empower consumers with the requisite information to make informed purchasing decisions.

Sector Wise Analysis

Transportation

The country has got fifth position in car sales over the world. This is obviously in a rising trend with rising income and urbanization which in turn yield more demand for fuel. India has only 0.6% EV (Electric Vehicles) sales over all the segments together. National mission for electric mobility was set up in the year 2011 intended to endorse electric motor vehicle and hybrid manufacturing. FAME (Faster Adoption and Manufacturing of (Hybrid &) Electric Vehicles) scheme India (National scheme on electric mobility) to subsidize EV and hybrid vehicles is another step from the government towards curbing carbon emission and its positive move towards renewable energy. As per the data available 32 crores kg of CO_2 reduction has been achieved by moving on to EV/hybrid vehicles across the country, thereby saving 52,000 L of fuel per day. Bharath stage VI norms on emissions, implemented in 2020, is another important milestone towards reducing the carbon credit.

The Indian railway system stands as the world's fourth-largest in terms of rail track length, trailing only China in rail passenger activity. Projected growth for this sector surpasses that of any other country, anticipating a threefold increase by the year 2050. Approximately half of India's conventional rail tracks are electrified, while the construction of its first high-speed line is still ongoing. Notably, rail transport accounts for one-third of India's land freight, a significant proportion by global standards, with coal being the predominant commodity. India's climate pledge includes a plan to augment the share of railways in total land transport from 36 to 45%, achieved through the development of dedicated freight corridors.

The Indian Railways organization aspires to attain the goal of Net Zero Carbon Emissions for non-traction energy consumption by the year 2030, thereby contributing to sustainable environmental practices. Non-Traction end-uses account for over 2,100 GWh of electricity use per year, projected to increase by 30% by 2030 as IR network and infrastructure expands. They have made action plans such as compliance of all new buildings to meet BEE standards and labeling. Cloud based centralized web portal for energy usage data monitoring and management, intelligent water management system at stations, energy efficiency in equipment and appliances such as BEE 5 star based appliances, BLDC (BrushLess Direct Current) fans and LED (Light Emitting Diode) lighting wherever possible, and last but not the least periodic monitoring and training to the staff.

Oil and Gas

The Republic of India holds a position of significance in the global oil market, consuming approximately 29% of its total energy supply from this crucial resource in the year 2017. Marked by a rapid and substantial increase, India's oil demand is projected to potentially exceed double its current level by the year 2040. Such a surge could propel India to surpass the People's Republic of China as the leading contributor to global oil demand growth. However, it is noteworthy that India's domestic oil production has experienced a declining trend. In an effort to mitigate the impact of rising oil prices, the government of India has implemented substantial subsidies for the consumption of oil and gas. Nevertheless, between the years 2014 and 2016, these subsidies underwent a significant reduction of approximately 75%, resulting in a decrease to $6.8 billion. This reduction can be attributed to a combination of implemented reforms and the prevailing decline in oil prices. Furthermore, it is understood that the government is engaged in the pursuit of additional reforms in this domain.

In light of the urgent need to mitigate climate change, India has placed a high priority on implementing subsidy reductions and tax augmentations pertaining to petroleum and diesel. A notable initiative in this regard is the "Give it Up" campaign, which has been launched by the Indian government. The objective of this campaign is to encourage financially well-off citizens to voluntarily forgo their LPG cooking gas subsidy. Nonetheless, it is projected that the price of kerosene, which is widely utilized in India for lighting and cooking purposes, might experience a substantial increase, potentially disproportionately affecting individuals who lack access to electricity. Furthermore, the government's significant investment in road construction is expected to result in an elevated consumption of diesel. It is worth noting that natural gas has not made a significant contribution to India's energy mix, accounting for merely 6% of energy consumption in 2017. At present, India relies on imports to meet approximately half of its gas requirements, primarily sourcing gas from countries such as Qatar, the United States, Australia, and Russia.

Agriculture and Forests

Within the geopolitical entity of India, the agricultural sector, a noteworthy contributor to greenhouse gas emissions, holds accountability for approximately sixteen percent of the total emissions. Of this, methane from livestock and rice cultivation is the biggest contributor, at 74%. Nitrous oxide from fertilizers makes up the remaining 26%.

Within the territorial boundaries of India, agriculture serves as a significant provider of employment opportunities, with an estimated two-thirds of the population relying on farming activities for their sustenance. Agriculture, therefore, represents a substantial source of employment within the nation. Notably, India possesses an impressive 15% of the global cattle population, comprising approximately 300 million cows and buffaloes as of 2014. Additionally, India makes a significant contribution to the global milk production, accounting for approximately 19% of the total worldwide output.

To meet the demands of its expanding populace, India must significantly boost its production of food grains, as outlined in its climate commitment. However, the nation is already experiencing substantial climate variability, characterized by frequent droughts and floods. In order to reduce its greenhouse gas emissions and meet its climate goals, India will need to find ways to make its agriculture sector more sustainable. This could include measures such as improving irrigation efficiency, reducing the use of fertilizers, and switching to more climate-resilient crops.

India is taking steps to reduce agricultural emissions and improve food security. They're promoting lower-methane rice production, crop diversification, chemical-free farming, and soil health pilot projects. They also made it compulsory to coat urea with neem in 2015 to reduce nitrous oxide emissions. India is also installing solar water pumps to reduce energy consumption in agriculture. They've already installed 200,000, and they plan to install another 2.5 million.

In recent years, India has made considerable progress in augmenting its forest cover and aspires to achieve even more significant expansion. The nation aims to increase its forest cover to approximately 109 million hectares, a substantial increase from the 24% attained in 2013, with the goal of achieving a 33% forest cover. Additionally, India aims to absorb between 2,500 and 3,000 million tonnes of carbon dioxide (CO_2) by 2030 through additional forest and tree cover.

A key initiative, the Green India Mission, plays a pivotal role in facilitating India's achievement of its ambitious targets. Within a decade, it endeavors to extend tree cover by 5 million hectares and enhance the quality of an additional 5 million hectares of existing cover. Furthermore, the Indian government is offering incentives to encourage state-level efforts towards expanding forest cover, linking it to funding allocations.

It is noteworthy that India is undertaking impressive measures to mitigate its environmental impact and enhance food security. Monitoring the progress of these programs in the coming years will be of great interest.

Coal

India is the world's second largest consumer of coal, after China. India's coal use has been increasing rapidly, and it is expected to continue to do so in the coming years. This growth is being driven by India's rapid economic development and its need for electricity. Coal currently generates 76% of India's electricity.

India currently possesses 221 gigawatts (GW) of operational coal-based power plants, constituting the third largest such fleet in the world. Additionally, another 36GW is presently under construction, with a further 58GW in earlier developmental stages. However, it is worth noting that the pipeline for new plants is diminishing rapidly. The government's current projections indicate a coal capacity of 238GW by the year 2027. Nonetheless, some experts argue that this capacity could potentially become "stranded assets" if India does not abandon its plans to construct new coal-fired plants. According to CAT (Climate Analytics), abandoning plans to build new coal-fired plants could render India's policies compatible with the 1.5 °C (1.5 C) climate target.

The future of coal in India is uncertain, with some experts predicting that new coal plants The uncertain future of fossil fuels can be attributed to reduced electricity demand, exceeding expectations, and the declining cost of renewable energy. There is also concern about the health impact of coal plants, as air pollution from these plants is a major cause of death in India. The declining costs of solar energy and battery storage are substantially influencing the coal sector. Notably, in 2015, India established stricter emission standards for coal-powered plants, yet these standards remain unfulfilled.

Being the second-largest coal producer and importer, India falls second only to China in terms of coal production and consumption. Coal India, the state-owned coal mining enterprise of the country, dominates domestic output, constituting approximately eighty-four percent. With proven coal reserves estimated at ninety-eight billion metric tons (accounting for nine-point-five percent of the world's total), India, in this regard, also ranks second only to China. India relies on coal to maintain a reliable and affordable electricity supply. However, the government aims to eliminate coal imports presently sourced from nations such as Indonesia, South Africa, and Australia. Additionally, India has executed measures to enhance the efficiency of its coal-based power plants.

In the year 2010, the nation of India implemented a levy (tax) known as a coal "cess." This initiative generated approximately $12 billion in revenue between the years 2010 and 2018. Nevertheless, subsidies provided to the coal industry remain substantial, amounting to $2.3 billion in the year 2016. These subsidies are primarily realized through tax incentives granted to the mining sector.

In conclusion, India holds a significant position as a major participant within the international coal market. The country possesses extensive coal reserves and stands as the second-largest coal producer globally. The government of India is actively pursuing measures to reduce coal imports while simultaneously striving to enhance the efficiency of its coal-powered plants. Despite these efforts, coal continues to constitute a primary source of energy in India, and subsidies allocated to the coal sector remain elevated.

Figure 1 provides a visual representation of the diversification of greenhouse gas emissions across various sectors in India. It presents the distribution of total greenhouse gases, quantified in tonnes of carbon dioxide equivalents, emitted by each sector. The chart demonstrates that electricity and heat production contribute significantly to emissions in comparison to other sectors. Agriculture, followed by manufacturing and construction (primarily cement and related materials), and transportation subsequently rank in decreasing order of contribution to greenhouse gas emissions. Land use change and forestry has experienced the least change over the period under consideration. There is a dip towards 2019 in the top two sectors due to the increased awareness of global warming and its consequences. This is due to the positive effect created by the implementation of various policies by the government.

Figure 2 depicts the GHG emissions per person per year for the year 2019 alone, which gives the clear picture on the major contributors.

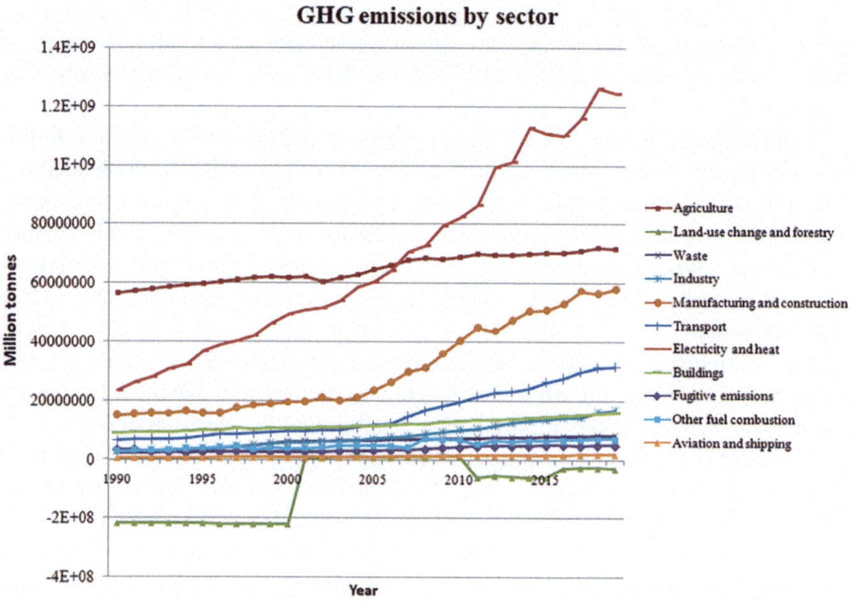

Fig. 1 Greenhouse gas emissions by sector in India in the period 1990–2019 [4]

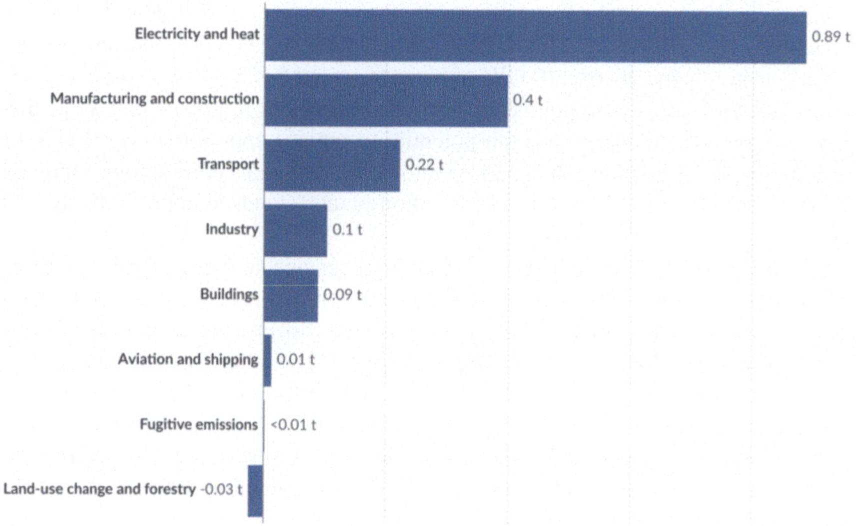

Fig. 2 Per capita greenhouse gas emissions by sector in India for the year 2019 [4]

Harnessing Renewable Energy Potential in India

India has made great strides in expanding its renewable energy sector in recent years. In 2017, India invested more in renewable energy than in fossil fuels for the first time. However, renewable energy still only accounts for a small percentage of India's electricity mix, at 15.5% in 2017. Of this, 9.1% came from large hydropower projects [5].

The precipitous decline in the cost of solar photovoltaic (PV) panels has diminished the appeal of coal-fired power for distributors of electricity. Notwithstanding, as the proportion of renewable energy sources within the grid continues to expand, novel policy measures will be required to accommodate the variable supply characteristics inherent in renewable energy sources. For example, smart grids can help to balance the grid by storing energy when there is excess supply and releasing it when there is demand.

India's ambitious renewable energy targets are a positive step towards a cleaner and more sustainable future. However, it is important to ensure that these targets are met in a way that does not compromise the reliability of the electricity grid.

In the year 2014, India's subsidies for renewable energy amounted to 431 million US dollars. However, by the year 2016, these subsidies had experienced substantial growth, increasing by a factor of six to reach 1.4 billion US dollars. Nevertheless, it is noteworthy that subsidies for traditional energy sources such as coal, oil, and gas continued to surpass those allocated for renewable energy, reaching a level six times higher in the year 2016.

India's promotion of renewable energy can be traced back to the year 2003, marked by the enactment of the Electricity Act. Subsequently, in the year 2015, India established an ambitious goal of installing 175 gigawatts (GW) of renewable energy capacity by the year 2022. This target encompasses 100 GW of solar energy, 60 GW of onshore wind energy, 10 GW of bioenergy, and 5 GW of small hydropower. According to a guest post published by Carbon Brief in the preceding year, this growth in renewable energy has the potential to replace approximately 70 GW of coal-fired power generation. It is pertinent to note that India's current total capacity, encompassing both fossil fuel and low-carbon plants, stands at approximately 350 GW.

In order to facilitate the widespread adoption of renewable energy, the government of India has implemented a range of measures. These measures encompass mandating power distributors and major electricity consumers to procure a portion of their energy from renewable sources, exempting solar and wind energy from inter-state transmission fees, and launching the National Solar Mission in the year 2010. The Jawaharlal Nehru National Solar Mission (JNNSM), also commonly referred to as the National Solar Mission, represents a joint initiative undertaken by the government of India and state governments, with the objective of promoting solar energy within India.

As of August 2018, India had installed or commissioned 23 gigawatts (GW) of solar photovoltaic capacity, as indicated in its recent biennial report to the United

Nations Framework Convention on Climate Change (UNFCCC). Additionally, the country is engaged in the construction of several of the world's largest solar parks. To facilitate large-scale grid-connected solar power projects, India has implemented a program titled "Development of Solar Parks and Ultra Mega Solar Power Projects" with an established target capacity of 40 GW by March 2024.

Solar Parks provide a streamlined and standardized approach for solar power developers by offering essential infrastructure, including land, power evacuation facilities, road connectivity, and water supply, along with all necessary statutory clearances. As of October 31, 2022, a total of 56 Solar Parks have been approved with a cumulative capacity of 39.28 GW across 14 states. Solar power projects with an aggregate capacity exceeding 10 GW have already been commissioned in 17 parks, and the remaining parks are currently in various stages of implementation. During the period from January to October 2022, solar projects totaling 832 MW were successfully commissioned in various Solar Parks.

India stands as the world's fourth-largest producer of wind energy, following China, the United States, and Germany. Remarkably, onshore wind capacity has experienced a threefold increase over the past decade, while solar energy prices have witnessed a significant decline of two-thirds since 2014.

India possesses abundant potential for leveraging renewable energy sources, with an estimated 1,100 gigawatts (GW) of commercially viable resources, encompassing 300 GW of wind power and 750 GW of solar power. The Climate Policy Initiative (CPI) asserts that by 2030, India could effectively integrate 390 GW of cost-efficient wind and solar generation into its electricity grid. This integration would represent a significant growth compared to the current level of 100 GW of renewable energy generation in India. Incorporating more renewable energy sources into the grid offers numerous advantages, such as reducing reliance on fossil fuels, enhancing air quality, and fostering employment opportunities within the renewable energy sector.

However, to facilitate the seamless integration of more renewable energy into the grid, certain challenges need to be addressed. These include the necessity to construct new transmission lines and storage facilities, as well as ensuring that the grid possesses the capacity to accommodate the intermittent nature of renewable energy sources.

The Ministry of New and Renewable Energy (MNRE), the governing body for renewable energy policy development in India, encompasses the National Institute of Solar Energy and the National Institute for Wind Energy. These institutions concentrate on research and development, testing, certification, standardization and skill development centered around renewable energy. The MNRE also includes energy from waste, biogas, bioenergy for electricity, and the Indian Renewable Energy Development Agency (IREDA), which operates as a non-banking financial institution providing loans for renewable energy and energy efficiency projects.

India's climate pledge acknowledges that a significant portion of its population relies on inefficient and polluting biomass energy, aiming to generate 10 gigawatts (GW) of cleaner, more efficient bioenergy by 2022, with 9 gigawatts (GW) already achieved in 2018.

In the hydropower sector, India possesses 4.5 gigawatts (GW) of small hydropower plants below 25 megawatts (MW) and aims to reach 5 gigawatts (GW) by 2022. The country's hydropower capacity, including large plants, expanded from 30 gigawatts (GW) in 2005 to 45 gigawatts (GW) in 2018, with a dedication to proactively pursue the development of its substantial hydropower potential.

In India, the government deems nuclear power a dependable, environmentally sustainable, and economically viable energy source, as outlined in its 2015 climate commitment. While the nation currently possesses 6.8 GW of nuclear capacity, the objective is to enhance this by ninefold to 63 GW by the year 2032. Nevertheless, a continuous debate persists concerning the advantages and disadvantages of nuclear energy within the country.

Additionally, India possesses one of the most extensive thorium reserves globally, considered a safer alternative to traditional nuclear fuels. India has consistently demonstrated a keen interest in experimental thorium reactors. However, widespread thorium-based power generation remains improbable before the 2050s, if ever.

In accordance with the Prime Minister's declaration at COP26, the Ministry of New and Renewable Energy is actively pursuing the goal of achieving 500 GW of installed electricity capacity from non-fossil sources by 2030. As of 31.10.2022, India has successfully installed a total of 172.72 GW of capacity from non-fossil fuel sources. This encompasses 119.09 GW of Renewable Energy, 46.85 GW of Large Hydro, and 6.78 GW of Nuclear Power capacity. As of 31.10.2022, this represents 42.26% of the total installed generation capacity of 408.71 GW within the country [6].

India has made significant progress in the development of renewable energy in recent years. As of 2022, India ranks 4th in the world in terms of installed renewable energy capacity, including large hydropower. It is also 4th in the world in terms of wind power capacity and solar power capacity. In the first 10 months of 2022, India added 14.21 GW of renewable energy capacity, up from 11.9 GW in the same period of 2021. During the same period, India generated 151.94 BU (Billion Units—*one unit being 1 kWh*) of electricity from renewable sources, up from 128.95 BU in 2021. This growth is due to a number of factors, including government support for renewable energy, falling costs of renewable technologies, and increasing awareness of the environmental benefits of renewable energy. India's progress in renewable energy is a positive development for the country and the world. Renewable energy is a clean and sustainable source of energy that can help to reduce greenhouse gas emissions and address climate change. India's continued investment in renewable energy will help to make the country a leader in the global transition to a clean energy future.

On November 2, 2022, the National Bioenergy Program, which includes the Waste to Energy Program, Biomass Program, and Biogas Program, was officially launched. The following are the various Bioenergy schemes that the Ministry is currently implementing:

- Program on Energy from Urban, Industrial, and Agricultural Wastes/Residues.

- Scheme to Support Promotion of Manufacturing of Briquettes & Pellets and Biomass (Non-Bagasse) Based Cogeneration in Industries in the Country for the period from Fiscal Year (FY) 2021–22 to 2025–26.
- Biogas Power (Off-Grid) Generation and Thermal Application Program (BPGTP).
- New National Biogas and Organic Manure Programme (NNBOMP).

As of the year 2022, the aggregate installed capacity of biomass power and cogeneration projects approximated 9.4 gigawatts, whilst the capacity of waste-to-energy projects stood at 223.14 megawatts. Additionally, a significant augmentation of 61.12 megawatt equivalents was witnessed in the installed capacity of waste-to-energy projects during the period commencing on 1st January 2022 and concluding on 31st October 2022.

A pivotal emphasis has been placed on the development of human resources, with initiatives such as the *Vayumitra* and *Jalmitra* Skill Development Programmes being implemented under the auspices of the Human Resource Development Scheme. Furthermore, equity infusion amounting to Rs. 1500 crores has been made in IREDA, and Rs. 1000 crores in SECI. The *Jal-Urjamitra* Skill Development Programme has the objective of training technicians in the maintenance of Small Hydro Power Projects.

In a bid to ensure energy and water security and generate supplementary income for farmers, the government has launched the PM-KUSUM Scheme. This scheme endeavors to augment solar capacity by 30.8 gigawatts with the provision of central financial support.

The implementation process for rooftop solar schemes has been streamlined by developing a National Portal, through which consumers can apply directly, obviating the need to await the finalization of tenders or empanelment of vendors by Discoms. Since its inception, the portal has received applications for a solar capacity of 117 megawatts, of which projects aggregating over 18 megawatts have been deemed feasible.

The National Green Hydrogen Mission has been conceived with the aim of positioning India as a hub for green hydrogen production and export. The Ministry of New and Renewable Energy (MNRE) is spearheading the development of this mission, with the objectives of decarbonizing major economic sectors, achieving energy independence, and establishing India as a paradigm for the global clean energy transition.

Conclusion

Positioned as a critical participant in the global energy market, India's population of 1.4 billion and rapidly expanding economy are key factors. The government's achievements in providing electricity, promoting clean cooking practices, implementing energy market reforms, and deploying renewable energy sources are all part of a comprehensive strategy. The government's vision is to ensure secure, affordable, and sustainable energy access for all citizens. Between 2000 and 2019, around 750 million people in India gained access to electricity, demonstrating effective policy implementation. This chapter explores strategies for achieving these goals, emphasizing energy system transformation, security, and affordability. Additionally, it highlights valuable lessons from India's energy sector that can be applied by other countries.

India's energy sector is a complex and dynamic one, with a wide range of stakeholders and interests. The government has a central role in shaping the sector, but it also works with a variety of private and public sector partners. The government's energy policy is focused on ensuring energy security, affordability, and sustainability. To achieve these goals, the government has implemented a number of reforms, including:

- Liberalizing the energy market
- Investing in renewable energy
- Promoting energy efficiency.

These reforms have had a positive impact on the energy sector, but there are still challenges to be addressed. India's energy demand is growing rapidly, and the country is still heavily reliant on fossil fuels. The government needs to continue to invest in renewable energy and energy efficiency in order to meet its energy goals.

In the context of global efforts to mitigate climate change, India's per capita carbon dioxide emissions currently stand at 1.6 tonnes, significantly below the global average of 4.4 tonnes. Notwithstanding this, India contributes approximately 6.4% to global carbon dioxide emissions. Recognizing the gravity of the situation, India actively engages in international forums dedicated to combating climate change. India's Nationally Determined Contribution (NDC) under the Paris Agreement delineates specific objectives to reduce the emissions intensity of its economy, enhance the share of non-fossil fuels in its power generation capacity, and create an additional carbon sink through the expansion of forest and tree cover. While India has made commendable progress in reducing the emissions intensity of its GDP in accordance with its plans, achieving a low-carbon electricity supply remains a significant challenge.

The following are some of the challenges that India faces in achieving its NDC targets:

- India's economy is growing rapidly, which is leading to increased energy demand.
- India's power sector is still heavily reliant on fossil fuels, such as coal.
- India has a large population, which makes it difficult to implement large-scale energy efficiency measures.
- India's infrastructure is not well-developed, which makes it difficult to connect remote areas to the electricity grid.

Despite encountering challenges, India is making significant progress in reducing its greenhouse gas emissions. The administration has implemented a range of policies and initiatives to promote renewable energy and energy efficiency. Additionally, India is investing heavily in research and development to create innovative technologies that will support the achievement of its Nationally Determined Contribution (NDC) targets.

India's dedication to renewable energy, especially solar power, has led to substantial advancements. Notably, India's solar PV investments have surpassed those in all fossil fuel-based electricity generation sources combined. This remarkable accomplishment demonstrates India's commitment to reducing its reliance on fossil fuels and addressing climate change. India has also undertaken commendable steps towards enhancing energy efficiency, such as investing in energy efficiency certificates, which have yielded impressive results.

India's vulnerability to climate change impacts, such as water stress, storms, floods, and extreme weather events, is severe. This underscores the critical need for the government to prioritize adaptation and resilience in the energy sector. Furthermore, given that the energy sector is a significant water consumer, the government must ensure that energy planning takes into account the water-energy nexus and future space cooling requirements as India's energy demand continues to grow.

India's energy demand is projected to double by 2040. Increasing energy efficiency goals could save India $190 billion and 875 terawatt hours of electricity per year by 2040. This is significant because it would help India meet its energy requirements without having to add significant power generation capacity. However, India's energy sector is a valuable resource for other countries. The government's experience in reforming the energy market and developing renewable energy could be helpful to other countries that are facing similar challenges.

References

1. IPCC, 2018: Summary for Policymakers. In: Global Warming of 1.5°C, available on https://www.ipcc.ch/site/assets/uploads/sites/2/2019/06/SR15_Full_Report_Low_Res.pdf [Online Resource]
2. Jocelyn Timperley (2019)—"The Carbon Brief Profile: India", available on https://www.carbonbrief.org/the-carbon-brief-profile-india/
3. Impact of energy efficiency measures for the year 2019–20 available on https://beeindia.gov.in/sites/default/files/publications/files/BEE_Final%20Report_Website%20version.pdf
4. Hannah Ritchie, Pablo Rosado and Max Roser (2020)—"Emissions by sector" Published online at OurWorldInData.org. Retrieved from: 'https://ourworldindata.org/emissions-by-sector' [Online Resource]
5. India 2020—Energy Policy Review, International Energy Agency.
6. Year-End Review 2022-Ministry of New and Renewable Energy available on https://pib.gov.in/PressReleasePage.aspx?PRID=1885147

Emission Abatement: A Cop28 Outlook for Countries

Fotouh Al-Ragom

Abstract The issue of climate change has been a global concern for several decades, and the need to address greenhouse gas emissions has become increasingly urgent. The annual Conference of the Parties (COP) organized under the United Nations Framework Convention on Climate Change (UNFCCC) plays a central role in international efforts to combat climate change. COP28, the 28th installment of this critical conference, holds the promise of advancing global efforts to mitigate emissions and limit the devastating impacts of climate change. This essay will provide an in-depth exploration of the concept of emission abatement and its significance in the context of COP28. It will examine the current state of global emissions, the key challenges and opportunities countries face in abating emissions, and the potential outcomes and agreements that may emerge from COP28. The essay will also discuss the role of various stakeholders, including governments, businesses, and civil society, in driving emission abatement efforts. By the end, readers should have a comprehensive understanding of the complexities and nuances of emission abatement in the context of COP28.

Keywords Climate change · Greenhouse gases · Temperature · Environmental concerns · Barriers

Introduction

COP28, the first of its kind in the Middle East—a region severely exposed to the effects of climate change—took place in Dubai in the United Arab Emirates (UAE) and proved to be a momentous occasion. The main goal of the conference, which took place from November 30 to December 13, 2023, was to evaluate the status of the Paris Agreement, a landmark agreement signed in 2015, and to develop plans to drastically cut greenhouse gas emissions. One of the most significant results of COP28 was the

F. Al-Ragom (✉)
Energy and Building Research Center, Kuwait Institute for Scientific Research, Kuwait City, Kuwait
e-mail: fragom@kisr.edu.kw

K. J. Sreekanth (ed.), *Energy Efficiency Improvements with Emission Abatement for Energy Sustainability*, https://doi.org/10.1007/978-981-97-9480-5_4

creation of a loss and damage fund, which was an essential reaction to the dire effects of climate change on developing countries. However, the pledged $700 million falls notably short of the estimated annual cost of damages, reflecting a persistent gap in addressing the global south's demands. Another ground breaking achievement was the adoption of a fossil fuel phase-out agreement, signalling a commitment to transition away from coal, oil, and gas, aiming for net-zero emissions by 2050. This was a landmark move, marking the first explicit acknowledgment of the need to end fossil fuel use. COP28 also witnessed the launch of a $30 billion private market climate capital, poised to mobilize private sector investment in low-carbon and climate-resilient projects, especially in emerging and developing economies. This initiative, backed by the World Bank, the International Finance Corporation, and the World Economic Forum, aligns with efforts to leverage private resources in the pursuit of sustainable solutions.

The subsequent sections examine the critical aspects of emission abatement, offering an in-depth analysis of the current state of global emissions with a focus on major contributors like carbon dioxide, nitrous oxide, and methane. Challenges and opportunities in emissions abatement are also discussed.

The key themes of the 28th Conference of the Parties to the UN Framework Convention on Climate Change (COP28) are discussed in this chapter, along with the important role that the conference played in establishing the global agenda for emission reduction.

The subsequent sections will also scrutinize how nations utilize this forum to strengthen their emission reduction goals by observing the intricate negotiations and commitment revisions occurring throughout COP28.

Assessing progress becomes a focal point, allowing a collective evaluation of whether global efforts align with the urgency of limiting temperature increases. The ethical dimensions of emission reduction are dissected in the context of equity and climate justice, shedding light on strategies to rectify the disproportionate impacts on vulnerable populations. COP28's discussions on climate finance take centre stage, providing insights into commitments and financial mechanisms crucial for supporting mitigation and adaptation projects, particularly in developing nations.

Furthermore, the dynamic landscape of stakeholder engagement will be showcased, illustrating the active roles played by businesses, civil society, and non-governmental organizations in the pursuit of emission abatement initiatives. The unique contributions of various stakeholders from governments, businesses, civil society, scientists, and individuals will be outlined, to highlight that the collaborative and inclusive nature is essential for comprehensively addressing climate change on a global scale.

The influence of research on climate change will be also discussed emphasizing key fields that significantly contribute to its impact. Moreover, technological innovations and solutions that can propel the achievement of climate goals will be elaborated on.

Current State of Global Emissions

In an era where environmental concerns are on the rise, comprehending the difficulties and urgency involved in cutting emissions becomes even more crucial. It's critical to understand the current status of global emissions to properly appreciate these challenges. Significant amounts of greenhouse gases, mostly carbon dioxide (CO_2), methane (CH_4), and nitrous oxide (N_2O), have been released into the atmosphere as a result of human activity, primarily the burning of fossil fuels and deforestation. These gases cause climate change by trapping heat, which raises Earth's temperature. It becomes clear when we examine the challenges associated with emission abatement that treating the underlying causes of these emissions is essential to the health of our planet and its people. The subsequent sections will delve into different types of emissions.

Greenhouse Gas Emissions

Global greenhouse gas emissions continue to rise, primarily driven by the burning of fossil fuels for energy, transportation, and industrial processes. As of the latest available data, in 2021, the global Green House Gases (GHG) CO_2 equivalent emissions reached 54.59 billion metric tonnes, a significant increase from previous years. This upward trend in emissions is a cause for concern, as it exacerbates the climate crisis. The major contributors to this increase are China, the United States, and India. Figure 1 shows the greenhouse gas emissions of the world and major contributing countries. China's emissions are due to intensive manufacturing. As per Wu and Han's findings, China's direct greenhouse gas (GHG) emissions are predominantly driven by the manufacturing sector. In 2016, this sector accounted for 64.39% of the total direct GHG emissions, amounting to 6,598.44 million metric tonnes of CO_2 equivalent. Notably, 90.00% of the carbon dioxide (CO_2) emissions within this sector were attributed to energy consumption [1].

The global map and annual CO_2 emissions are shown in Fig. 2 [2]. China released 12.38 billion tonnes of CO_2 in 2022, compared to 5.2 billion tonnes in the US in

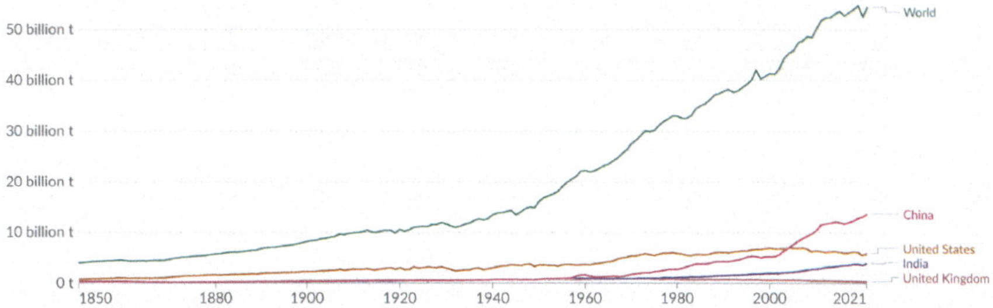

Fig. 1 Greenhouse gas emissions (CO_2 eq.) [2]

the same year. India and Russia emitted 2.86 billion tonnes of CO_2 and 1.97 billion tonnes of CO_2 respectively.

The main emitting sector is the electricity and heat as shown in Fig. 3, followed by transport and then manufacturing and construction [3]. In 2020, the share of the various sectors in CO_2 emissions is shown in Fig. 4 [3]. The electricity and heat share were 44% followed by transport (20% then manufacturing and construction (18%).

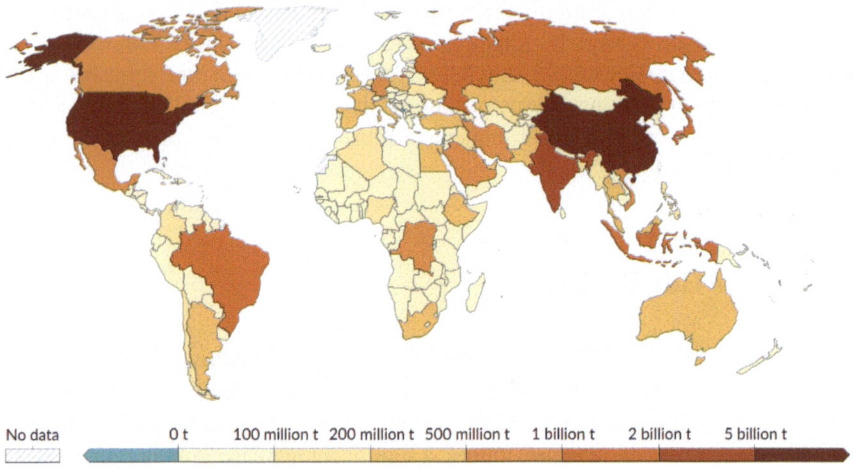

Fig. 2 Annual CO_2 emissions including land-use change, 2022 [2]

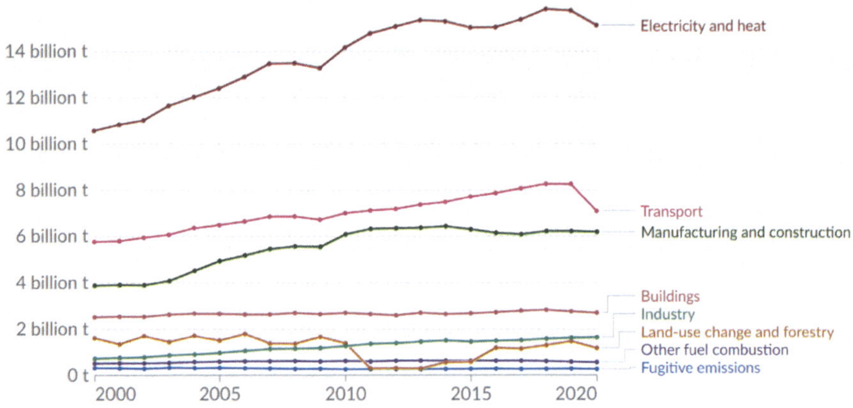

Fig. 3 World CO_2 emissions by sector [3]

Fig. 4 CO_2 emissions share by sector

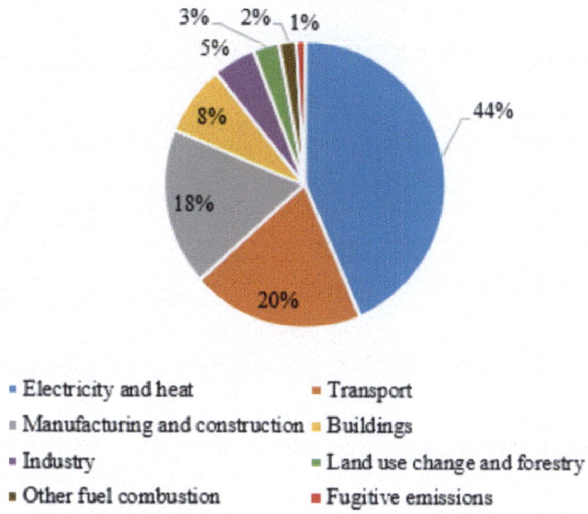

The Role of Methane and Nitrous Oxide

To measure the effect on global warming, the Global Warming Potential (GWP) is used. This metric is used to estimate how much each type of greenhouse gas (GHG) contributes to global warming over a given amount of time, typically 100 years. The amount of heat that a greenhouse gas retains in the atmosphere in relation to carbon dioxide (CO_2), which is frequently used as a reference, is measured. The higher the GWP value, the more effective the gas is at trapping heat and contributing to global warming. CO_2 is assigned a GWP of 1. Methane (CH_4) and nitrous oxide (N_2O), while present in lower concentrations than CO_2, have much higher GWP values. Methane has a GWP of over 25 times that of CO_2 over a 100-year period, making it a potent but relatively short-lived greenhouse gas. Nitrous oxide, on the other hand, has a GWP nearly 300 times that of CO_2 over the same timeframe.

Methane emissions are produced by a variety of processes, such as burning biomass in wildfires, cattle digestion, agriculture, and natural gas production. Methane is also emitted by biological processes in wastewater treatment facilities, due to anaerobic conditions in rice fields, and during the breakdown of organic waste in landfills. Methane emissions in the energy production industry are a result of the extraction, processing, and transportation of fossil fuels. Methane can also be produced by the flooding of wetlands, particularly in tropical areas. This demonstrates the intricate interactions between man-made and natural processes in methane formation as they relate to waste management, energy production, agriculture, and environmental factors. The various sectors contributing to methane emissions are shown in Fig. 5. China was the major methane emitting country in 2021 with 1.81 billion tonnes of CO_2 eq. India, USA, Brazil and Russia follow China with a total methane emission of 897.01, 671.88, 588.7, 486.67 million tonnes of CO_2 eq. respectively [3].

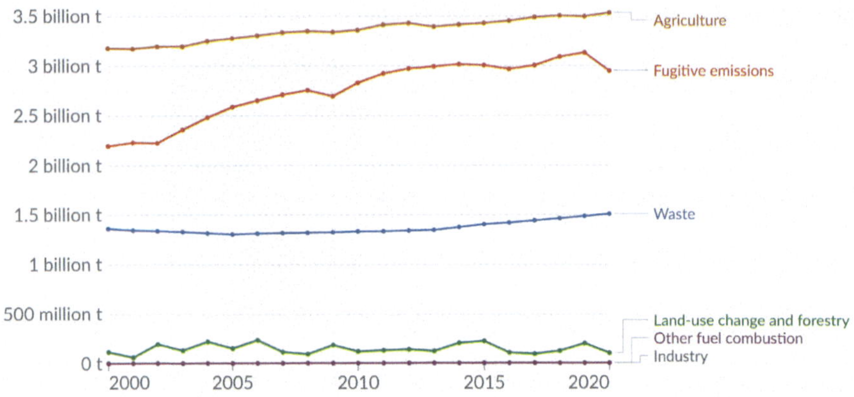

Fig. 5 Methane sector-wise emissions [3]

One of the most potent greenhouse gases, nitrous oxide, comes from a variety of sources, most notably from agricultural practices including manure management and applying fertilizer. Industrial processes, such as the production of nitric acid and combustion activities, also contribute to nitrous oxide emissions. Plants that treat wastewater using biological processes are also capable of producing this greenhouse gas. The wide range of intricate sources highlights the necessity of all-encompassing approaches, such as improved industrial operations and sustainable farming methods, to lessen the substantial influence of nitrous oxide on climate change. According to ourworld in data, a world map illustrating the nitrous oxide emissions in tonnes of CO_2 equivalent in 2021 is shown in Fig. 6 [2].

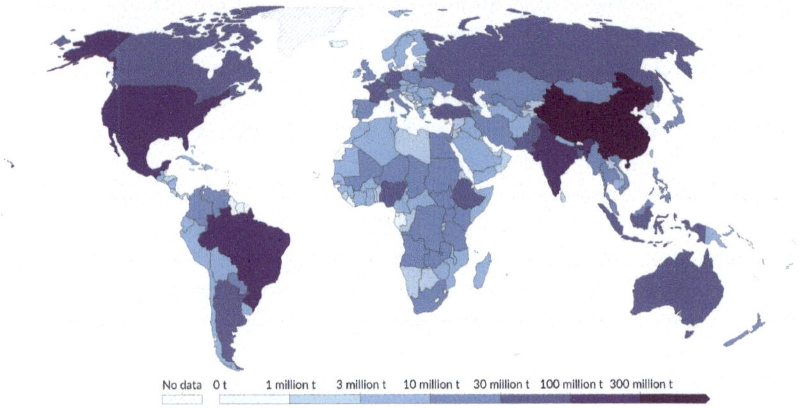

Fig. 6 Nitrous oxide emissions in 2021 [2]

Challenges and Opportunities in Emission Abatement

Reducing greenhouse gas emissions, or emission abatement, is a difficult task with many facets. While there are many obstacles that nations must overcome in their attempts to reduce emissions, there are also many chances to make significant strides in this direction.

Challenges in Emission Abatement

Economic Interests and Constraints

Economic interests and constraints significantly shape the challenges nations face in reducing greenhouse gas emissions. Many countries heavily depend on industries that are major emitters, like fossil fuel extraction, making them resistant to swift changes due to vested economic interests. Additionally, emission reduction strategies often demand substantial investments in new technologies and infrastructure, posing economic challenges that are particularly pronounced for nations with limited resources. The tension between economic considerations and the imperative for sustainable practices underscores the complexity of navigating emission reduction on a global scale.

Technological and Infrastructural Barriers

Navigating technological and infrastructural barriers stands as a significant challenge in the transition toward cleaner energy sources and sustainable practices. Embracing these changes often requires extensive technological and infrastructural transformations, which, in turn, can incur substantial costs and consume considerable time. While advancements in clean energy technologies are pivotal for emission reduction, challenges such as high initial costs and limited scalability present hurdles to their widespread adoption. The need for extensive updates to existing systems, along with the development and implementation of innovative technologies, calls for strategic planning and investment. Overcoming these barriers demands a concerted effort to foster research, development, and widespread deployment of sustainable technologies, ensuring that the transition to cleaner energy is not only effective but also economically viable and accessible on a global scale.

Political Resistance

International agreements can be difficult to negotiate, particularly when dealing with nations that place a higher priority on economic growth than environmental protection. Climate change can be a divisive political topic. Strict emission reduction regulations may be difficult to implement due to political concerns and opposition to change, especially in countries where some industries have strongholds. An example is the Conference of the Parties (COP), that is the international body tasked with addressing climate change. COP struggles to unite global efforts and encourage coordinated action. Bridging the divergent interests and priorities of member nations remains a significant challenge, impeding the formation of cohesive strategies for mitigating climate change on a unified front.

Equity and Justice

In the context of emission reductions, equity and justice present additional difficulties. When it comes to allocating responsibility for mitigation activities, the historical context makes clear that industrialized countries have made considerable contributions to greenhouse gas emissions. This creates a complex dynamic. Despite having been significant historical emitters, some countries might be reluctant to shoulder a fair part of the responsibility, which might spark discussions about accountability and fairness in international climate negotiations. The international community must carefully balance honoring historical contributions with establishing an appropriate allocation of emission reduction duties to secure a just and equitable transition to a sustainable future for all countries. In the fight against climate change, the pursuit of justice and equality highlights the need for inclusive, adaptable policies that take into consideration the various socioeconomic conditions of different countries worldwide.

Global Coordination

To achieve significant emission reductions, global coordination is essential, and worldwide collaboration is required to meet the common challenge of climate change. Nonetheless, the absence of a universal agreement among countries may hinder the development and execution of successful international policies. Aligning countries towards a cohesive approach is complicated by divergent agendas, different economic factors, and diverse levels of development. The lack of a unified worldwide commitment could lead to dispersed initiatives, which would make it difficult to meet collective emission reduction targets and lessen the overall effects of climate change. Finding common ground and exercising diplomatic grace are necessary to bridge these divides, as it is acknowledged that concerted efforts are necessary to produce long-lasting and significant results on a worldwide basis. worldwide forums, like the Conference of the Parties (COP), play a crucial role in fostering dialogue and cooperation, yet the ongoing challenges highlight the need for continuous and concerted efforts to build consensus for effective climate action [4].

Dependency on Fossil Fuels

Fossil fuel dependence continues to be a major roadblock in the effort to reduce emissions. The problem is not just the environmental effects of using fossil fuels; it's also the complex web of economic and infrastructural ties that have grown over time. It is important to overcome ingrained systems and habits to make the shift to renewable energy sources. Examples of these include financial investments, job arrangements linked to the fossil fuel industry, and current energy infrastructures. Adopting new technology is just one part of this shift; other social and economic issues related to dependency on fossil fuels must also be addressed. Due to the size and complexity of the problem, comprehensive approaches that consider the financial effects on communities, businesses, and countries are needed to ensure a fair and long-lasting shift away from reliance on fossil fuels. Despite the growing awareness of the need for cleaner alternatives, the process of disentangling from fossil fuels remains a multifaceted undertaking, demanding coordinated efforts from governments, industries, and communities to successfully navigate this pivotal shift in the global energy landscape.

Opportunities in Emission Abatement

Renewable Energy (RE)

Renewable Energy (RE) emerges as a pivotal solution in the quest for emission reduction. The rapid advancement of RE technologies, exemplified by the tremendous progress in solar and wind power, presents a transformative opportunity to shift away from fossil fuels and significantly mitigate greenhouse gas emissions. Solar and wind power technologies have witnessed remarkable improvements in efficiency, cost-effectiveness, and scalability, making them increasingly viable alternatives to traditional energy sources. The expansion of renewable energy promotes innovation and economic growth in addition to making the energy landscape greener and more sustainable. Global adoption of these technologies has the potential to change the energy paradigm and offer a vital route to a low-carbon future. To fully realize the potential of renewable energy in reducing emissions and promoting a sustainable energy transition, however, issues like intermittency, storage, and the requirement for supporting infrastructure must be successfully resolved.

Sustainable Agriculture

Implementing sustainable agricultural practices is crucial for mitigating climate change, as it can effectively reduce methane and nitrous oxide emissions, two potent greenhouse gases associated with conventional farming methods.

Methane and Nitrous Oxides are emitted in agriculture practices. Thus to mitigate the generated emissions some technologies, techniques and practices are available. As an example, the use of GPS-guided tractors and drones as a precision farming technique can optimize the use of fertilizers and pesticides, minimizing environmental impact. In the USA, where agriculture energy input accounts for 17% to the total consumed energy, precision agriculture has been adapted. This has led to reduction of wasteful use of agricultural inputs on the environment while simultaneously improving crop output and lowering input costs through the optimal application of fertilizer, chemicals, seeds, and irrigation resources. The use of GPS-based auto guidance system reduced overlapping of agriculture equipment and vehicles passes, hence saving fuel and lessening the environmental impact [5]. Unmanned aerial vehicles (UAVs), remotely piloted aircraft systems (RPAS), and normalized difference vegetation index (NDVI) approaches were tested in small farms along the Portuguese-Spanish border. Through the use of study assessment techniques, they have discovered that significant productivity factor savings are made possible by an effective UAV/RPAS and NDVI combination, supporting sustainable agriculture [6].

Other techniques can also be utilized to enable sustainable agriculture including agroforestry, incorporating trees and shrubs into farming systems, enhances carbon sequestration while providing biodiversity benefits. Cover cropping, another sustainable practice, involves planting crops like legumes during off-seasons to enhance soil fertility, reducing the need for synthetic fertilizers and curbing nitrous oxide emissions. By embracing these practices and others like organic farming and rotational grazing, the agricultural sector can contribute significantly to emission abatement, promoting a more environmentally responsible and resilient food production system.

International Cooperation

COP conferences provide a platform for international cooperation and agreement on emission reduction targets, offering an opportunity to create a global consensus on the issue. The Conference of the Parties (COP) conferences serve as pivotal platforms for international cooperation among nations to address climate change and establish collective commitments toward emission reduction targets. These gatherings are held under the United Nations Framework Convention on Climate Change (UNFCCC) and serve as crucial forums fostering international cooperation and negotiating agreements on emission reduction targets [4].

A noteworthy instance is the 2015 COP21 Paris Agreement, which committed 196 nations to keeping rises in global temperatures well below 2 degrees Celsius over pre-industrial levels. These conferences' collaborative atmosphere promotes information exchange, technological transfer, and financial aid from industrialized to underdeveloped countries. To further demonstrate worldwide cooperation, the Global Environment Facility (GEF) is a financing instrument that supports projects and programs targeted at combating climate change, biodiversity loss, and land degradation [7]. A common commitment to halting climate change and realizing sustainable development objectives globally is fostered in large part by the COP conferences.

The most recent COP28, that was held in Dubai, UAE, more than 100 countries agreed to transition away from "Fossil Fuels". The text "The beginning of the end of fossil fuels" [8–10].

Technological Innovation

The importance of technological innovation in mitigating greenhouse gas emissions is crucial in tackling climate change. Technological developments in carbon capture and storage, for example, make it possible to capture and sequester carbon dioxide emissions from power production and industrial activities, thereby limiting their atmospheric release. Emission reduction can also be greatly aided by the adoption of energy-efficient techniques across a range of businesses. A more sustainable and low-carbon future is made possible, for instance, by the development and broad use of energy-efficient industrial techniques, smart grids, and renewable energy technology. Continuous technical progress, including innovations in energy storage systems derived from renewable sources or the application of artificial intelligence to optimize energy usage, offers prospects to improve efficiency and lower emissions in several industries. By harnessing the power of innovation, nations and industries can transition towards cleaner and more sustainable practices, fostering a collective effort to combat climate change. Policy and Regulation: Implementing and enforcing strong environmental policies and regulations create a framework for emission reduction, encouraging industries to adopt cleaner practices. As an example of technology innovation, the United States (U.S.) Department of Energy/National Energy Technology Laboratory (DOE/NETL) Carbon Capture Program is working towards the development of the next generation of advanced CO_2 capture concepts for the power-generation sector. They targeted the development of several technologies utilizing solvents, sorbents, membranes and novel concepts for carbon capture such as water gas shift (WGS) WGS/CO_2 separation process and Hybrid systems [11].

Public Awareness and Engagement

To catalyze climate change action, public engagement, and awareness campaigns can play a major role as they can cultivate the demand for sustainable products and services. As individuals become more informed about the environmental consequences of their choices, there is a growing expectation for businesses to prioritize emission reduction and adopt eco-friendly practices. For example, because of consumers' growing demand for items with reduced carbon footprints, eco-friendly certifications and labels are becoming more popular. In response to this need, businesses are integrating sustainable practices including using green technologies, adopting circular economy ideas, and using renewable energy sources into their operations. Furthermore, community-based initiatives and online platforms create channels for collective activism, amplifying public voices and holding businesses responsible for their environmental footprint. Noteworthy campaigns, including those championing divestment from fossil fuels, have gained broad attention, shaping corporate conduct. Likewise, movements like supporting local businesses "Buy Local" and adopting "Zero-Waste" practices urge individuals to make environmentally conscious choices, thereby diminishing their overall impact on the environment [12, 13].

Outreach and education initiatives help raise public awareness by enabling people to make knowledgeable decisions and igniting a shared commitment to addressing climate change. Businesses are seeing the significance of integrating emission reduction methods into their core principles as public opinion toward sustainability goals grows. By doing so, they not only meet customer expectations but also support a more robust and sustainable global economy [14].

International Collaboration

International cooperation is a vital part of the global effort to reduce emissions because it allows countries to work together and trade technologies, resources, and expertise. Globally focused collaborative efforts addressing climate change have the potential to combine special talents and global discoveries. This information exchange procedure expedites the development and implementation of effective emission reduction programs and enables a more equitable allocation of resources and responsibilities. Initiatives such as joint research projects, technology transfer agreements, and cooperative funding techniques are examples of collaborative efforts to combat climate change. As nations work together to address this shared issue, the synergy of many perspectives and capacities becomes a driving force toward attaining meaningful and sustainable development. Many cooperative initiatives comprise the Sustainable technology transfer between developed and poor countries is a common area of collaboration. Sustainable technology transfer between developed and poor countries is a common area of collaboration. For example, the Kyoto Protocol's Clean Development Mechanism (CDM) makes technology transfer and sustainable development initiatives in underdeveloped nations easier. International cooperative initiatives (ICIs) are cooperative projects with the potential to cut greenhouse gas emissions that involve several nations and non-state players. To provide crucial coordination between ICIs and national government-led efforts to mitigate climate change, the United Nations Environment Programme (UNEP) is required [14].

Circular Economy Practices

Circular economy practices stand as a compelling avenue for emission reduction by embracing principles that prioritize the reuse and recycling of materials [15]. The adoption of circular economy principles offers opportunities to mitigate emissions associated with both resource extraction and waste generation. By shifting away from traditional linear models of production and consumption, where resources are extracted, used, and then discarded, circular economies emphasize the continual reuse, refurbishment, and recycling of materials. This approach not only conserves valuable resources but also reduces the need for energy-intensive processes related to raw material extraction and production. Additionally, minimizing waste through recycling and repurposing contributes to a significant decrease in emissions linked to landfill disposal. Embracing circular economy practices, therefore, not only fosters sustainability but also provides a comprehensive strategy for addressing climate change by curbing emissions throughout the entire lifecycle of products. As businesses and communities increasingly adopt these principles, the potential for emission reduction grows, offering a transformative approach to sustainable resource management [16].

Addressing the challenges and leveraging available opportunities requires a comprehensive and integrated approach, involving governments, businesses, and the public. Successful emission abatement strategies often involve a mix of policy measures, technological innovation, and shifts in societal behavior.

The Role of COP28 in Emission Abatement

Addressing the varied effects of greenhouse gas (GHG) emissions on the earth is a major responsibility of the Conference of the Parties (COP) to the United Nations Framework Convention on Climate Change (UNFCCC) [17, 18]. COP gatherings, like COP28, offer a worldwide forum for countries to discuss and put into practice plans to lessen the effects of climate change. COP takes the following significant steps to mitigate the effects of GHG emissions:

1. Setting Emission Reduction Targets: Setting national goals for reducing emissions is made possible in large part by COP gatherings. One such example is the Paris Agreement, which was enacted during COP21 [19]. The Paris Agreement was adopted by 196 countries on December 12, 2015. The agreement attempts it establishes a global target to keep the average global temperature rise to well below 2 °C above pre-industrial levels and to limit the increase to 1.5° [17, 20].
2. Nationally Determined Contributions (NDCs): During COP meetings, countries submit their NDCs, outlining their commitments and actions to reduce GHG emissions. These contributions form a critical part of the global effort to address climate change. The UNFCCC established the Nationally Determined Contributions (NDC) Registry to serve as a platform that tracks and displays the commitments made by individual countries as part of the Paris Agreement [21]. Each country's NDC outlines its specific climate action plans, including emission reduction targets, adaptation strategies, and other contributions to mitigate climate change. The registry provides transparency and accountability, allowing stakeholders to monitor progress and assess the collective efforts of nations in addressing climate change.
3. Adaptation Strategies: Adaptation strategies are a vital component of COP discussions, acknowledging the reality that certain effects of climate change are already inevitable. These discussions revolve around supporting communities and ecosystems vulnerable to the impacts of changing climate conditions. The focus is on developing and implementing strategies that help these regions and groups adapt to the challenges posed by climate change. This might involve creating infrastructure resilient to extreme weather events, implementing agricultural practices suited to changing conditions, and supporting communities in managing water resources in the face of shifting rainfall patterns. The adaptation strategies discussed at COP meetings aim to enhance resilience and minimize the adverse impacts of climate change on societies and environments worldwide [22].

The adaptation policy cycle within the UN Climate Change regime involves a systematic and comprehensive approach to addressing the impacts of climate change is shown in Fig. 7. The process begins with a thorough assessment of the impacts, vulnerabilities, and risks associated with changing climate conditions in a specific region or community. This scientific evaluation serves as the foundation for developing adaptation plans, which outline targeted strategies and measures across various sectors. Implementation of these plans involves the execution of specific measures, such as infrastructure projects and the adoption of climate-resilient practices.it also provides political space for engagement. Ongoing monitoring and evaluation ensure the effectiveness of these measures, allowing for adjustments as needed to enhance resilience.

The adaptation policy cycle under the UN Climate Change Regime is a strategic framework designed to guide nations in addressing the impacts of climate change (Fig. 7). Beginning with the planning phase, countries are encouraged to develop

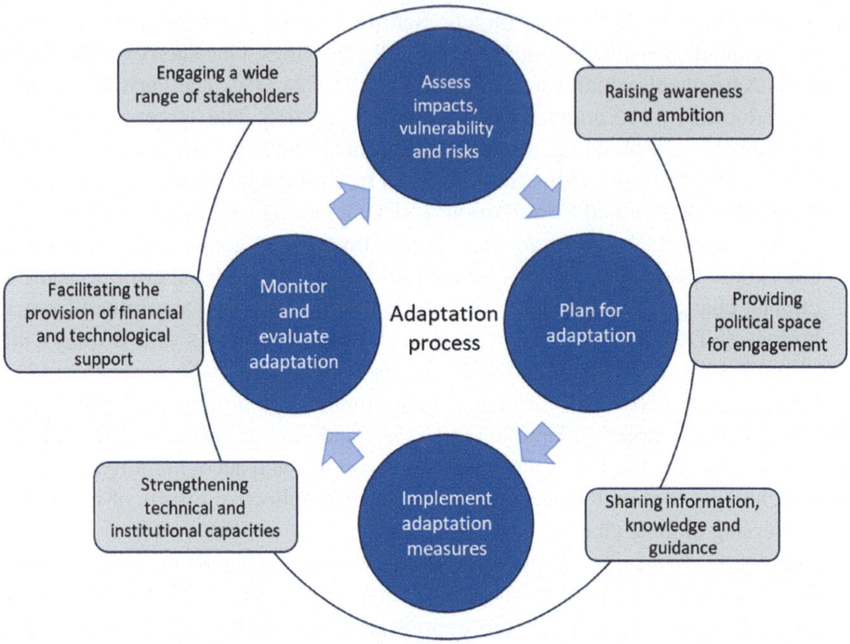

Fig. 7 The adaptation policy cycle and support offered under the UN Climate Change regime [23]

adaptation plans, fostering political engagement to create a supportive environment. Subsequently, the implementation phase involves sharing crucial information, knowledge, and guidance to facilitate the execution of specific adaptation measures. Continuous monitoring and evaluation are facilitated by the UN, ensuring nations receive necessary financial and technological support. The engagement of diverse stakeholders in assessing impacts, vulnerability, and risks enhances the effectiveness of adaptive planning. This comprehensive cycle enables countries to systematically navigate the challenges of climate change, with UN support reinforcing global commitment to adaptation and resilience.

4. Loss and Damage Mechanism: COP meetings deliberate on the Loss and Damage Mechanism, a crucial concept addressing the adverse effects of climate change. This mechanism involves discussions on providing financial support to countries and communities that experience losses beyond their capacity to adapt. Essentially, it recognizes that some impacts of climate change cannot be avoided, and affected regions may incur losses and damages that go beyond their ability to cope or recover. The Loss and Damage Mechanism emphasizes the need for assistance and compensation to support these vulnerable entities. For instance, if a small island nation faces severe damage from a climate-related event, the mechanism seeks to ensure that it receives the necessary financial aid to recover and build resilience against future climate impacts. This approach underscores

the commitment to fairness and solidarity in addressing the global challenges posed by climate change [24].

5. Promoting Climate Finance: Encouraging climate financing is one of the main goals of COP, highlighting the need for funding, especially for developing countries, to support adaptation and mitigation initiatives. The effective execution of initiatives meant to lower emissions and increase resilience depends on adequate finance. The creation of climate-resilient infrastructure and the shift to sustainable behaviors are made easier by this support. For instance, the UNFCCC established the Green Climate Fund (GCF) as a financial vehicle to support developing nations with their climate-related initiatives [25]. It helps countries achieve their targets for reducing emissions and building climate resilience by allocating cash to programs ranging from sustainable agriculture practices to renewable energy projects [26]. The global commitment to fair and efficient climate action is shown in the advocacy of climate financing at COP by guaranteeing that financial resources are available where they are needed most.
6. International Collaboration: International collaboration is a cornerstone of COP efforts to address climate change [27]. The conferences provide a platform for countries to work together on various fronts, including research, technology transfer, and capacity-building initiatives. Through collaborative endeavors, nations pool their resources, knowledge, and expertise to collectively tackle the complex and global challenge of greenhouse gas (GHG) emissions. This collaboration extends to sharing best practices, developing joint projects, and fostering a supportive environment for the implementation of climate-related policies. For example, countries may collaborate on the development and deployment of renewable energy technologies, share insights into successful emission reduction strategies, or engage in joint efforts to address cross-border environmental challenges. The emphasis on international collaboration at COP underscores the interconnected nature of climate change and the recognition that effective solutions require coordinated efforts on a global scale.
7. Monitoring and Reporting: Monitoring and reporting mechanisms established by COP play a crucial role in tracking the progress of countries toward meeting their emission reduction targets. These mechanisms are designed to ensure transparency, accountability, and adherence to the commitments made by nations. Countries are required to regularly report their greenhouse gas emissions and the actions they have taken to mitigate climate change. This not only provides a clear picture of each country's efforts but also facilitates the identification of areas that may need additional attention and support. Examples of monitoring and reporting initiatives include national greenhouse gas inventories, which detail emissions from various sectors, and periodic reviews conducted to assess a country's overall progress [28]. These measures contribute to a collective understanding of global efforts in emission reduction and hold nations accountable for their actions.
8. Engaging Stakeholders: Engaging stakeholders is a pivotal aspect of COP conferences, transcending traditional government-focused negotiations. The inclusion of businesses, civil society, and non-governmental organizations (NGOs) reflects a recognition that emission abatement requires a collective effort from diverse

sectors. Businesses play a significant role by showcasing their commitment to sustainability through innovative practices and technologies. For instance, multinational corporations may present initiatives involving renewable energy adoption, carbon–neutral operations, or circular economy practices. Civil society and NGOs contribute by highlighting grassroots initiatives, advocating for climate justice, and emphasizing the importance of community-driven solutions. Notable examples include environmental organizations promoting reforestation projects or indigenous groups championing sustainable land management practices. This inclusive approach amplifies the impact of emission abatement initiatives, fostering collaboration among governments, businesses, and civil society to address the complexities of climate change on a global scale.

Through the above-mentioned actions, COP contributes significantly to realizing the effects of GHG emissions and advancing a coordinated global response to mitigate climate change impacts.

COP28 Goals

COP28, during its convening, pursued a set of well-defined goals to address the urgent challenges posed by climate change. One primary objective was the expeditious advancement of the energy transition, recognizing the critical role of transitioning to sustainable energy sources in mitigating climate impacts. The conference aimed to rectify and strengthen climate finance mechanisms, acknowledging the crucial need for financial support to drive effective mitigation and adaptation projects globally. Additionally, COP28 prioritized placing nature, people, lives, and livelihoods at the forefront of climate action, emphasizing the interconnectedness of environmental and human well-being. Underpinning all endeavors was the overarching principle of full inclusivity, recognizing the importance of involving diverse stakeholders in shaping comprehensive and equitable climate solutions. These goals collectively emphasized the commitment of COP28 to fostering a sustainable and resilient future through transformative actions on a global scale.

The focus of COP28 was on following topics:

- A global stock-take: to offer a detailed evaluation of nations' combined efforts and advancements in combating climate change.
- The mitigation work program: to emphasize methods for cutting greenhouse gas emissions and make the shift to a low-carbon future.
- The global goal on adaptation: to acknowledge the need to improve resilience and adaptive ability in the face of climate impacts.
- Climate finance: to facilitate climate action and discuss financial plans designed to compensate for losses and damages.

COP28 Agenda

The opening ceremony was held on November 30, 2023ar Duabi Expo. On December 1, 2023 the Opening World Climate Action Summit was held followed by High-Level Roundtables and High-Level Events.

G77 & China Summit High-Level Roundtables High-Level Events Closing World Climate Action Summit. On December 2, 2023, G77 & China Summit, High-Level Roundtables, High-Level Events and Closing World Climate Action Summit were held. On these first three days mandated events were held and negotiation sessions. On the following days, the schedule was divided into Week 1 Thursday 30 November—Thursday 7 December and Week 2—Wednesday 6 December—Tuesday 12 December [29]. The last two days were dedicated to COP closing. During week 1 and 2, thematic days were held, with a different theme on each day. The themes were as follows:

- Health/relief, recovery and peace
- Finance/trade/gender equality/accountability
- Energy, industry, and just transition
- Multilevel action urbanization and built environment/transport
- Youth, children, education and skills
- Nature, land use, and oceans
- Food, agriculture and water.

Major Outcomes of COP28

The important outcomes included to transition away from fossil fuels, about 200 countries agreed on that. Additionally, the global renewable energy and energy efficiency pledge that call for tripling the world installed renewable energy generation capacity to 11,000 GW by 2030. It also called for doubling the rate of energy efficiency enhancements from 2 to 4% by 2030 [30]. The following Fig. 8 highlights the major outcomes of COP28.

Emission abatement stands as a global imperative, with the 28th COP conference serving as a pivotal moment in the ongoing endeavor to tackle climate change. The current status of global emissions, the obstacles and prospects in emission reduction, the significance of COP28, and the engagement of diverse stakeholders collectively chart the course for addressing the crisis. The results of COP28 carry the potential to establish the framework for future international efforts in emission reduction, influencing global actions to combat climate change and alleviating its severe consequences. It is imperative for nations to unite with determination, acknowledging that the future of our planet hinges on the actions taken today.

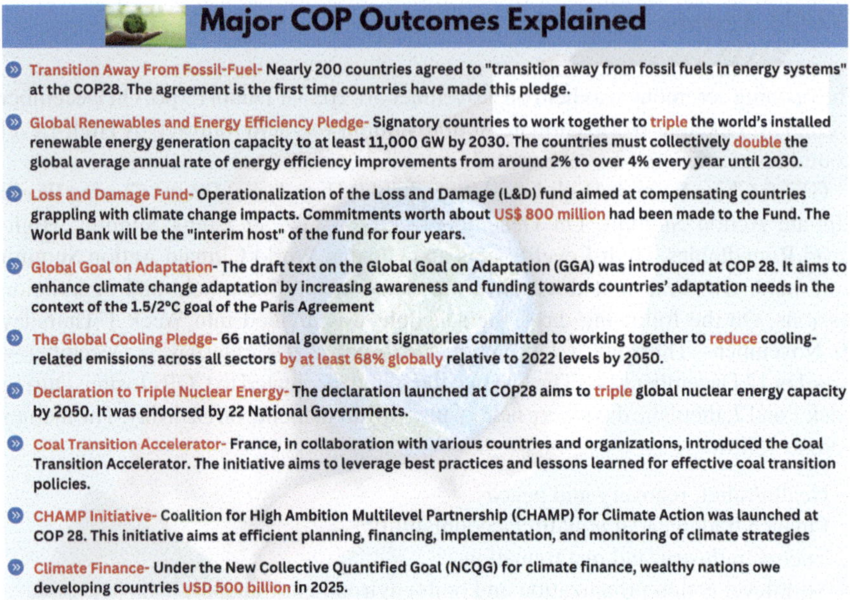

Fig. 8 Summary of the major COP28 outcomes [31]

The Impact of Research on Climate Change

Climate change has emerged as a major worldwide issue that needs immediate response. Its extensive impacts, which include warming temperatures, extreme weather, increasing sea levels, and threats to ecosystems and biodiversity, have a significant impact on the environment and on people. Scientific study and research are essential to understanding climate change and developing plans to reduce its effects and prepare for its aftermath. This section explores the crucial role that research plays in advancing efforts to address climate change, forming policy, and improving our understanding of the phenomenon. Research becomes a guiding force as we navigate the complexity of climate change, providing insights that direct sustainable behaviours, inform decision-making, and open the door for group action to resolve the issues raised by rapidly changing climate.

The Science of Climate Change

The Earth's climate system, its influencing factors, and the results of changes are examined in the wide subject of climate change research. Important insights into the causes, effects, and potential remedies of climate change are gained from research in this field. The relationship between weather, climate, and water-related sciences and services and the advancement of the Sustainable Development Goals (SDGs) is highlighted in Fig. 9. Various scientific topics and fields are related to climate change whether directly or indirectly.

Scientific research has unveiled the profound influence of the greenhouse effect on Earth's climate, emphasizing its natural role and the alarming consequences of human-induced intensification through activities like fossil fuel combustion and deforestation. Scientific research has been pivotal in establishing the anthropogenic impact on climate change, linking human activities to a surge in greenhouse gas emissions since the industrial revolution. Documented temperature increases, escalating extreme weather events, and rising sea levels are among the well-established findings, indicating the urgency of addressing climate change. The significance of scientific insights extends to shaping global climate policies, with institutions like the IPCC guiding decision-makers. Climate policy targets, as seen in the Paris Agreement, are rooted in research-backed understandings of temperature risks, while

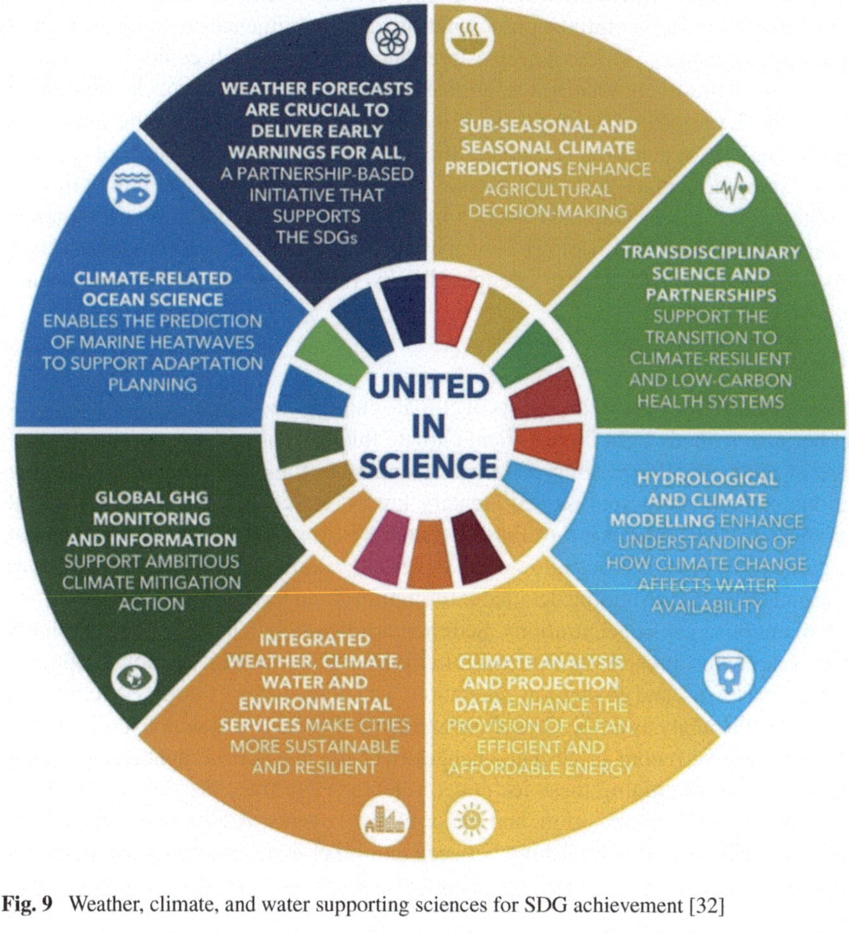

Fig. 9 Weather, climate, and water supporting sciences for SDG achievement [32]

national emission reduction commitments, carbon pricing mechanisms, and regulations are directly informed by scientific findings. Overall, scientific research not only elucidates the complexities of climate change but also underpins international efforts to combat its impact through informed decision-making and strategic policy measures. In this section, we will delve into specific scientific research topics that hold significance in the context of climate change.

Ecology and Biology

Research within Ecology and Biology is instrumental in comprehending the profound impact of climate change on the natural world. Scientists in these fields investigate how alterations in climate patterns reverberate through ecosystems, affecting biodiversity and reshaping the distribution of species. Ongoing research delves into the details of these transformations, exploring the responses of diverse organisms to shifts in temperature, precipitation, and habitat conditions. For instance, studies on migration patterns of birds or mammals provide insights into how species adapt to changing climates. Researchers also examine reproductive behaviour, studying how factors like altered temperature or precipitation regimes influence breeding cycles and population dynamics. Moreover, investigations into the overall ecological balance help uncover the interconnected relationships between different species within an ecosystem. Examples of such research include projects monitoring the impact of climate change on coral reefs, tracking the migration routes of endangered species, and studying how plant-pollinator interactions are influenced by changing climatic conditions. By interpreting the complex interplay between climate and ecosystems, research in Ecology and Biology contributes essential knowledge for conservation efforts and informs strategies to safeguard biodiversity in the face of ongoing climate challenges.

Geology

To comprehend the historical aspects of climate change, geological research is essential because it sheds light on historical climate fluctuations and informs our knowledge of current and emerging patterns. Scientists in the field of Geology employ various techniques to analyse sedimentary layers, ice cores, and other geological records, offering a glimpse into Earth's climatic history. For instance, the study of ice cores from polar regions allows researchers to examine air bubbles trapped over thousands of years, providing detailed information about past atmospheric conditions and greenhouse gas concentrations. Sedimentary layers in lakebeds or ocean floors serve as archives, preserving evidence of climate events such as volcanic eruptions, shifts in sea levels, and changes in vegetation. One noteworthy example of geological research is the analysis of the Greenland and Antarctic ice cores, which has provided a comprehensive record of climate changes over the last several hundred thousand years [33]. By examining these ice cores, scientists can trace the flow of temperatures, atmospheric composition, and major climate events. Additionally, geological studies contribute to distinguishing between natural climate variability and potential anthropogenic influences [34]. This historical context aids in identifying trends that may be attributable to human activities, such as the burning of fossil fuels and deforestation. Overall, research in Geology enriches our understanding of Earth's climate history, providing a valuable foundation for assessing the impacts of contemporary climate change and making informed decisions about our future.

Agriculture

In the field of agriculture, continuing scientific study is essential to comprehending and resolving the complex interplay between food production and climate change. The productivity of agriculture is significantly impacted by climate change, which is mostly evident in changes to temperature, precipitation patterns, and the frequency of extreme weather events. Scientists studying agriculture explore these issues and work to create crop varieties that are hardly enough to resist shifting climatic circumstances [35]. To ensure food security in the face of rising temperatures and water scarcity, for example, research on heat-tolerant crops or drought-resistant plant varieties is crucial.

Water management is a critical aspect addressed by research in response to changing climates. Scientists explore innovative irrigation techniques, water-efficient farming practices, and sustainable water resource management to optimize agricultural output while mitigating the impact on water availability [36]. Understanding how climate change affects pests and diseases is another facet of agricultural research, guiding the development of strategies for pest control and disease management in crops [37].

The focus of agricultural research is on sustainable farming methods. Research on precision farming, organic farming, and agroecology attempts to reduce agriculture's environmental impact while fostering climate variability resistance [5, 38, 39]. Research initiatives that evaluate the effects of cover crops on soil health and carbon sequestration, for instance, offer important new perspectives on sustainable agricultural methods [40].

Health Sciences

Within the field of Health Sciences, research plays a crucial role in comprehending the complex interrelationships between climate change and public health, revealing the various ways in which environmental modifications influence illness trends. Temperature, precipitation, and general environmental changes brought about by climate change affect disease prevalence and dispersion. Researchers studying public health conduct extensive research to identify these links and create practical plans for reducing the health hazards linked to climate change [41, 42].

The connection between temperature variations and the transmission of infectious diseases is a major topic of study. Research programs, for example, examine how rising temperatures help disease vectors—mosquitoes that spread diseases like malaria and dengue fever—expand their geographic range [43]. Comprehending these trends facilitates the creation of focused measures to manage illnesses transmitted by vectors.

Waterborne diseases are another critical aspect addressed by health research in the context of climate change. Alterations in precipitation patterns can lead to changes in water quality and the prevalence of waterborne pathogens. Public health scientists conduct investigations to assess the impact of climate-induced changes on waterborne diseases like cholera and develop strategies to ensure clean water access and sanitation.

Moreover, research in Health Sciences explores the broader implications of climate change on non-communicable diseases, mental health, and vulnerable populations. For example, studies may examine the relationship between extreme heat events and cardiovascular diseases or investigate the mental health impacts of climate-related disasters [44].

By uncovering these connections, public health researchers contribute valuable insights to the development of adaptive strategies, early warning systems, and public health interventions aimed at mitigating the health risks posed by climate change. The intersection of climate science and health research is crucial for building resilience within communities and healthcare systems to safeguard public health in a changing climate.

Engineering

In the field of engineering, continuing research is essential to addressing climate change concerns with creative ideas and cutting-edge technology. Engineers play a pivotal role in formulating and executing plans for adaptation and mitigation, thereby making a substantial contribution towards a sustainable future [45].

Research in renewable energy technologies is a cornerstone of engineering efforts in emission abatement [46]. Ongoing studies focus on enhancing the efficiency and affordability of solar, wind, and hydropower systems. For example, advancements in photovoltaic technology and energy storage solutions are crucial areas of exploration. Engineers are also involved in developing novel approaches to harnessing clean energy, such as wave and tidal energy technologies, to diversify the renewable energy portfolio [47, 48].

Sustainable infrastructure development is another key aspect of engineering research in emission abatement. This involves designing and implementing structures that are resilient to climate impacts while minimizing their carbon footprint. For instance, research may explore the use of green building materials, energy-efficient construction techniques, and smart infrastructure systems that optimize energy use [49–51].

To cut emissions from the automobile industry, engineers are studying and creating high-efficiency propulsion systems, alternative fuels, and electric and hybrid vehicles for the transportation sector. More sustainable urban mobility is also facilitated by the incorporation of smart transportation systems and technological developments in public transportation.

In the industrial sector, engineering research focuses on improving process efficiency, reducing energy consumption, and implementing carbon capture technologies. For example, studies may investigate ways to optimize manufacturing processes, develop cleaner industrial technologies, and implement circular economy principles to minimize waste [52].

These examples emphasize the interdisciplinary nature of engineering research in global warming abatement. Engineers play a crucial role in global initiatives to alleviate the impacts of climate change and advance a future that is both sustainable and resilient.

Social Sciences

Within the Social Sciences domain, research illustrates the far-reaching impacts of climate change on human societies. Social scientists focus on human behaviour in response to environmental risks, studying factors influencing climate change awareness and the adoption of sustainable practices. This knowledge is crucial for developing effective communication strategies and interventions that promote sustainable behaviours among individuals and communities.

Another key area of inquiry within social science research is the examination of migration patterns influenced by climate change. Scholars analyse how rising sea levels, extreme weather events, and shifting agricultural conditions contribute to displacement and migration [53]. This research informs policymakers on the complexities of climate-induced migration, guiding the formulation of humane and effective policies that address the rights and needs of affected populations [54]. Additionally, social scientists explore community resilience, investigating how different communities adapt to and recover from climate change impacts. The study of social networks, governance structures, and cultural practices informs the development of community-based adaptation initiatives, leveraging local knowledge and social capital to enhance resilience. To put it simply, social science research offers critical insights that influence policy, direct decision-making, and promote adaptive measures for a future that is more resilient and sustainable in the face of climate change.

Economics

In the field of Economics, researchers explore the profound economic implications of climate change, recognizing its potential to impact industries, resource availability, and overall economic stability [55]. To sustain economic health, it is imperative that climatic concerns be addressed. This multidisciplinary endeavour aims to comprehend the complex relationship between economic activity and environmental sustainability.

Economists play a pivotal role by evaluating the costs and benefits of various climate policies [56]. For instance, they assess the economic feasibility and impact of transitioning to renewable energy sources, analysing potential job creation and economic growth associated with such shifts. Additionally, economists explore market-based mechanisms like carbon pricing to internalize the external costs of greenhouse gas emissions, creating economic incentives for emission reduction.

Moreover, economists contribute to understanding the economic risks linked to climate change. Research in this area involves studying the potential costs of extreme weather events, sea-level rise, and disruptions to global supply chains. By analysing the financial vulnerabilities of industries heavily dependent on climate-sensitive resources, economists provide insights into how businesses and governments can adapt to mitigate economic risks associated with a changing climate. In essence, economic research offers crucial perspectives on navigating the intersection of climate change and economic systems, informing decision-making processes for a resilient and sustainable economic future [57].

The interdisciplinary nature of climate change research emphasizes the interconnectedness of scientific fields, illustrating how a comprehensive understanding of climate change requires collaboration across various disciplines to address its complex and multifaceted challenges.

Technological Innovations and Solutions

Technological innovations are at the forefront of the fight against climate change, driven by insights from climate change research. These innovations play a crucial role in both mitigating emissions and adapting to the challenges posed by a changing climate. Breakthroughs in renewable energy technologies, energy storage, and smart grids contribute to reducing carbon footprints and transitioning to cleaner energy sources. Additionally, advanced technologies in agriculture, climate modelling, and digital solutions enhance our ability to adapt to changing weather patterns and optimize resource use. Embracing and scaling these technological solutions is essential for building a more sustainable and resilient future in the face of climate change. Several technology innovations and solutions are discussed as follows.

Renewable Energy

Scientific research has been instrumental in boosting forward renewable energy technologies, notably in solar, wind, and hydroelectric power. Across the globe, countries have made significant strides in harnessing clean energy alternatives, showcasing the impact of research-led innovations [58, 59].

In the solar energy sector, countries like Germany have been pioneers in adopting photovoltaic technology. Germany's commitment to solar energy is evident in its extensive use of solar panels, particularly in residential and commercial settings. China, too, has emerged as a global leader in solar energy production, contributing significantly to the worldwide expansion of solar power capacity [60].

Wind power innovations have also made remarkable contributions to the renewable energy landscape. Denmark, for instance, has been a leader in wind energy development, with wind turbines contributing a substantial portion to its overall energy mix [61]. The United States, especially in states like Texas and Iowa, has embraced large-scale wind farms, making significant strides in wind energy production [62, 63].

In the domain of hydroelectric power, Norway stands out as a success story. Leveraging its abundant water resources, Norway has developed efficient and environmentally sustainable hydroelectric power plants, playing a key role in meeting its energy needs. Similarly, Brazil has made substantial investments in hydropower, with the Itaipu Dam being one of the largest and most impactful hydroelectric projects globally [64].

These examples highlight the global impact of scientific research on advancing renewable energy technologies and the pivotal role countries play in adopting and integrating these innovations into their energy portfolios.

Carbon Capture and Storage (CCS)

The technology breakthrough of Carbon Capture and Storage (CCS) is a notable example of how research has inspired technological innovation. With the use of this technique, carbon dioxide (CO_2) emissions from power plants and industrial activities are captured and safely stored underground to be released into the atmosphere. The necessity to control carbon emissions, a primary contributor to climate change, has propelled the development of CCS.

Nations globally are actively investigating and deploying Carbon Capture and Storage (CCS) initiatives to confront climate change and attain objectives for reducing emissions.

For example, since the 1990s, CO_2 from natural gas production has been captured and stored at Norway's Sleipner Project [65]. Norway also collaborated with Sweden in a project to capture CO_2 at Preem's Lysekil refinery in Sweden and transporting the captured CO_2 across borders to Norway [66, 67].

The United States is also utilizing CCS technology to capture emissions from coal-fired power plants through initiatives like Petra Nova in Texas [68]. In Canada, the Boundary Dam CCS Project in Saskatchewan is capturing CO_2 from a coal-fired power plant, demonstrating the technology's adaptability across diverse energy sources [69].

Electric Vehicles (EV)

Research activities have played a vital role in driving the advancement and broad adoption of Electric Vehicles (EVs). Compared to conventional internal combustion engine vehicles, these vehicles produce far fewer emissions, marking a revolutionary change in the automobile industry. Research-based EV development is an essential part of international programs to cut emissions, especially in the transportation sector.

Many countries are engaging in electric vehicle promotion to achieve sustainable transportation systems and lessen the burden of climate change. For instance, Norway has emerged as a leader in EV adoption, with a substantial percentage of its new car sales being electric [70].

Additionally, the Netherlands has implemented measures to phase out traditional gasoline and diesel vehicles in favour of electric alternatives, demonstrating a commitment to sustainable mobility [71].

These examples underscore the transformative impact of research on the global transition to electric vehicles, signalling a paradigm shift toward cleaner and more sustainable transportation solutions.

Energy Efficiency

Energy efficiency, backed by scientific studies, plays a vital role in emission reduction. Extensive research has been instrumental in driving the development of advanced technologies, appliances, and architectural designs geared toward optimizing energy consumption.

Countries worldwide are actively integrating energy-efficient solutions into their infrastructure to mitigate emissions. In Germany, energy efficiency has been promoted through initiatives such as "Energiewende," which focuses on transitioning to a more energy-efficient and sustainable future [72]. Japan has also made significant strides in energy efficiency, implementing stringent standards for appliances and promoting eco-friendly technologies [73]. Furthermore, the United States, through programs like ENERGY STAR, encourages the adoption of energy-efficient products and practices to reduce overall energy consumption [74].

These global examples highlight the direct correlation between research-driven advancements in energy efficiency and tangible efforts by nations to incorporate these solutions into their policies and practices, ultimately contributing to emission reduction on a broader scale.

Climate-Resilient Agriculture

Finding techniques for climate-resilient agriculture has been greatly aided by research, which has focused on creating practices that reduce emissions and adjust to changing climate conditions. Drought-resistant crop production and the use of sustainable farming methods are two of these tactics.

To improve food security and sustainability, numerous nations are incorporating climate-resilient agricultural practices. For instance, in India, the National Mission for Sustainable Agriculture (NMSA) program was launched that aims at promoting climate-robust farming methods such as the adoption of crop types that are resilient to climate change and effective water management [75]. Australia has used conservation tillage and precision farming techniques to improve agricultural resilience in the face of fluctuating and unpredictable weather [76, 77]. In addition, Brazil, a major force in world agriculture, has adopted sustainable land management techniques to lessen the effects of climate change on its farming industry [78].

These examples underscore the vital role of research in informing practical strategies for climate-resilient agriculture, with nations implementing these findings to bolster the resilience of their agricultural systems in the face of a changing climate.

Conclusions and Recommendations

Challenges in emission abatement are multifaceted, encompassing economic interests and constraints that often hinder swift transitions. Technological and infrastructural barriers pose significant hurdles, and political resistance may impede the implementation of stringent policies. Equity and justice concerns also add complexity, as historically developed countries grapple with their disproportionate contributions to emissions. The need for global coordination forms a crucial basis for effective emission reduction strategies. However, amidst these challenges lie opportunities for transformative change. Renewable energy stands out as a beacon, providing a sustainable alternative to traditional sources. Sustainable agriculture practices offer a dual benefit by reducing emissions and promoting environmental resilience. International cooperation becomes pivotal in sharing resources and expertise. Technological innovation emerges as a powerful tool, driving advancements in emission reduction technologies. Public awareness and engagement foster demand for sustainable practices, while international collaboration encourages a unified approach. Circular economy practices, emphasizing reuse and recycling, further contribute to the arsenal of opportunities for emission abatement. Through a comprehensive exploration of these challenges and opportunities, a nuanced understanding of the intricate landscape of emission abatement emerges.

COP, as a vital international forum, plays a crucial role in advancing emissions abatement strategies to safeguard our planet. Through collaborative negotiations, commitment updates, and the engagement of diverse stakeholders, COP contributes significantly to shaping global efforts in reducing greenhouse gas emissions. The outcomes of COP meetings influence international policies, commitments, and actions, serving as a linchpin in the collective endeavour to address climate change and mitigate its far-reaching impacts.

This chapter explores COP28 discussions, aiming to provide the reader with current outcomes to inspire action and accelerate progress in alleviating the challenges posed by climate change.

In conclusion, urgent action is paramount. Initiating efforts to meet the 2030 and 2050 targets, especially focusing on enhancing energy efficiency and expanding renewable energy capacities, is imperative. International collaboration becomes pivotal in uniting efforts and sharing knowledge. Utilizing available funding mechanisms can facilitate successful energy transitions in developing countries. The time to act is now, emphasizing the critical need for collective and immediate action to address climate change and achieve sustainable goals.

References

1. S. Wu and H. Han, "Sectoral changing patterns of China's green GDP considering climate change: An investigation based on the economic input-output life cycle assessment model," J. Clean. Prod., vol. 251, p. 119764, 2020, https://doi.org/10.1016/j.jclepro.2019.119764.

2. H. Ritchie, P. Rosado, and M. Roser, "CO_2 and Greenhouse Gas Emissions" 2023. Published online at OurWorldInData.org. https://ourworldindata.org/co2-and-greenhouse-gas-emissions (accessed Dec. 24, 2023).
3. H. Ritchie, P. Rosado, and M. Roser, "Emissions by sector: where do greenhouse gases come from?" 2020. Published online at OurWorldInData.org. https://ourworldindata.org/emissions-by-sector (accessed Dec. 24, 2023).
4. Conference of the Parties (COP). https://unfccc.int/process/bodies/supreme-bodies/conference-of-the-parties-cop
5. G. C. Bora, J. F. Nowatzki, and D. C. Roberts, "Energy savings by adopting precision agriculture in rural USA," Energy. Sustain. Soc., vol. 2, no. 1, pp. 1–5, 2012, https://doi.org/10.1186/2192-0567-2-22.
6. L. Loures, A. Chamizo, P. Ferreira, A. Loures, R. Castanho, and T. Panagopoulos. Assessing the effectiveness of precision agriculture management systems in mediterranean small farms. Sustainability, vol. 12, no. 9, p.3765, 2020.
7. Global Environment Facility (GEF). https://www.thegef.org/
8. COP28 Agreement Signals "Beginning of the End" of the Fossil Fuel Era. 2023. https://unfccc.int/news/cop28-agreement-signals-beginning-of-the-end-of-the-fossil-fuel-era (accessed Dec. 24, 2023)
9. Outcome of the first global stocktake. UNFCCC. https://unfccc.int/documents/636608 (accessed Dec. 24, 2023)
10. V. Volcovici, K. Abnett and J. Spring, "COP28 forced into overtime as fossil fuel phase-out divides countries." Reuters 2023. https://www.reuters.com/sustainability/climate-energy/cop28-presidency-wants-historic-mention-fossil-fuels-text-up-nations-2023-12-12/ (accessed Dec. 24, 2023)
11. L. Brickett, R. Munson, and J. Litynski, "U.S. DOE/NETL large pilot-scale testing of advanced carbon capture technologies," Fuel, vol. 268, May 2020, https://doi.org/10.1016/j.fuel.2020.117169.
12. J. R. Castigliego, A. Pollack, C. J. Cleveland, and M. J. Walsh, "Evaluating emissions reductions from zero waste strategies under dynamic conditions: A case study from Boston," Waste Manag., vol. 126, pp. 170–179, 2021, https://doi.org/10.1016/j.wasman.2021.02.026.
13. "Zero Waste Boston," 2023. https://www.boston.gov/environment-and-energy/zero-waste-boston (accessed Dec. 23, 2023).
14. A. Guaran and F. Venturini, "Zero Waste for geographical education on sustainability," Geogr. Notebooks, vol. 5, no. 1, 2022, https://doi.org/10.7358/gn-2022-001-guve.
15. A. Popović and V. Radivojević, "The circular economy: Principles, strategies and goals," Econ. Sustain. Dev., vol. 6, no. 1, pp. 45–56, 2022, https://doi.org/10.5937/esd2201045p.
16. Ferrovial, "Circular Economy," 2023. https://www.ferrovial.com/en/sustainability/environment/circular-economy/ (accessed Dec. 20, 2023).
17. P. Christoff, "The promissory note : COP 21 and the Paris Climate Agreement," Env. Polit., vol. 25, no. 5, pp. 765–787, 2016, https://doi.org/10.1080/09644016.2016.1191818.
18. N. Dacic, A. B. White, R. S. Ajimal, K. M. Boisvert, L. E. Oriol, and S. Akash, "How UNFCCC ' s COP Can Achieve Carbon Neutrality," J. Sci. Policy Gov., vol. 20, no. 1, 2022, https://doi.org/10.38126/JSPG200102.
19. D. Bodansky, "The Legal Character of the Paris Agreement," Rev. Eur. Comp. Int. Environ. Law, 2011, https://doi.org/10.2139/ssrn.2735252.
20. R. Falkner, "The Paris Agreement and the new logic of international climate politics," Int. Aff., vol. 5, no. July, pp. 1107–1125, 2016, https://doi.org/10.1111/1468-2346.12708.
21. UN Climate Change, "Nationally Determined Contributions Registry," 2023. https://unfccc.int/NDCREG?gclid=CjwKCAiAnL-sBhBnEiwAJRGigsk-q_ta_4gB7TrDV7htUPFkmrIo X1XnNl-DmKeQaevqhrWlXxJm9RoChwYQAvD_BwE (accessed Dec. 13, 2023).
22. LDC Expert Group, "The National Adaptation Plan Process," 2012. [Online]. Available: chrome-extension://efaidnbmnnnibpcajpcglclefindmkaj/https://unfccc.int/resource/docs/publications/publication_ldc_napp_2013.pdf.

23. UN Climate Change, "Adaptation and Resilience," 2023. https://unfccc.int/topics/adaptation-and-resilience/the-big-picture/introduction (accessed Dec. 22, 2023).
24. UNDP, "LOSS AND DAMAGE," UNDP Climate Promise, 2023. https://climatepromise.undp.org/what-we-do/areas-of-work/loss-and-damage?gad_source=1&gclid=CjwKCAiAnL-sBhBnEiwAJRGigtsCUS6xdklUEadi8qjPCco6YRrrxFolXpkFLUKbAi7tgsXTWCh4gRoCUmAQAvD_BwE (accessed Dec. 24, 2023).
25. GCF, "Green Climate Fund," 2023. https://www.greenclimate.fund/ (accessed Dec. 21, 2023).
26. C. Watson, N. Bird, L. Schalatek, and K. Keil, "Climate Finance Regional Briefing: Asia," Clim. Financ. Fundam., no. 8, pp. 1–4, 2017, [Online]. Available: http://www.climatefundsupdate.org/resources/finance-fundamentals.
27. United Nations Climate Change, "Call for Action: Partnerships for 2022 and 2023 including COP 27," 2023. https://unfccc.int/about-us/partnerships/current-calls-for-partnerships/call-for-action-partnerships-for-2022-and-2023-including-cop-27 (accessed Dec. 22, 2023).
28. C. Umemiya, "Improving GHG inventories by regional information exchange : a report from Asia," Carbon Balance Manag., vol. 4, pp. 7–10, 2006, https://doi.org/10.1186/1750-0680-1-9.
29. UNFCCC, "UN Climate Change Conference—Dubai Overview Schedule," no. December. Dubai, UAE, 2023, [Online]. Available: https://unfccc.int/documents/632476.
30. W. Obergassel et al., "From regime-building to implementation : Harnessing the UN climate conferences to drive climate action," WIREs Clim. Chang., no. June, pp. 1–12, 2022, https://doi.org/10.1002/wcc.797.
31. Forum IAS, "COP 28 Outcomes and Shortcomings—Explained Pointwise+ Infographics," 2023. https://forumias.com/blog/cop-28-outcomes-and-shortcomings-explained-pointwise-infographics/ (accessed Dec. 25, 2023).
32. L. Stuart, J. Luterbacher, R. Devillier, L. Paterson, K. Solazzo, and I. Bhasin, "United in Science," 2023. [Online]. Available: https://wmo.int/publication-series/united-science-2023.
33. C. Buizert, "The Ice Core Gas Age-Ice Age Difference as a Proxy for Surface Temperature Geophysical Research Letters," Adv. Earth Sp. Sience, pp. 1–12, 2021, https://doi.org/10.1029/2021GL094241.
34. G. Vasileios et al., "A 120 , 000-year long climate record from a NW-Greenland deep ice core at ultra-high resolution," Sci. Data, pp. 1–9, 2021, https://doi.org/10.1038/s41597-021-00916-9.
35. A. Ortiz-bobea, T. R. Ault, C. M. Carrillo, R. G. Chambers, and D. B. Lobell, "Anthropogenic climate change has slowed global agricultural productivity growth," Nat. Clim. Chang., vol. 11, no. April, 2021, https://doi.org/10.1038/s41558-021-01000-1.
36. A. Ahmad, J. Ordoñez, P. Cartujo, and V. Martos, "Remotely piloted aircraft (RPA) in agriculture: A pursuit of sustainability," Agronomy, vol. 11, no. 1, Jan. 2020, https://doi.org/10.3390/agronomy11010007.
37. S. Skendži, M. Zovko, and I. Paja˘, "The Impact of Climate Change on Agricultural Insect Pests," Insects, pp. 1–31, 2021, https://doi.org/10.3390/insects12050440.
38. L. Loures, A. Chamizo, P. Ferreira, A. Loures, R. Castanho, and T. Panagopoulos, "Assessing the effectiveness of precision agriculture management systems in mediterranean small farms," Sustainability, vol. 12, no. 9, May 2020, https://doi.org/10.3390/su12093765.
39. S. Rahman and A. Reza, "Productivity and efficiency impact of climate change and agroecology on Bangladesh agriculture," Land use policy, vol. 94, p. 104507, 2020, https://doi.org/10.1016/j.landusepol.2020.104507.
40. C. C. Okolie, G. Danso-abbeam, O. Groupson-paul, and A. A. Ogundeji, "Climate-Smart Agriculture Amidst Climate Change to Enhance Agricultural Production: A Bibliometric Analysis," Land, 2023, https://doi.org/10.3390/land12010050.
41. C. Linares, G. S. Martinez, V. Kendrovski, and J. Diaz, "A new integrative perspective on early warning systems for health in the context of climate change," Environ. Res., vol. 187, no. May, p. 109623, 2020, https://doi.org/10.1016/j.envres.2020.109623.
42. S. Schneider, A. Von Winning, F. Grüger, S. Anderer, R. Hoffner, and L. Anderson, "Physical Activity, Climate Change and Health — A Conceptual Model for Planning Public Health Action at the Organizational Level," Int. J. Environ. Res. Public Heal. Concept, 2022, https://doi.org/10.3390/ijerph19084664.

43. D. Campbell-lendrum and R. Reithinger, "Dengue and climate change," TRENDS Parasitol., vol. 18, no. 12, p. 4922, 2002.
44. W. L. Filho et al., "An analysis of climate change and health hazards: results from an international study," Int. J. Clim. Chang. Strateg. Manag., vol. 14, no. 4, pp. 375–398, 2022, https://doi.org/10.1108/IJCCSM-08-2021-0090.
45. M. Mier and C. Weissbart, "Power markets in transition: Decarbonization , energy efficiency , and short-term demand response," Energy Econ., vol. 86, p. 104644, 2020, https://doi.org/10.1016/j.eneco.2019.104644.
46. R. Janzen, M. Davis, and A. Kumar, "Greenhouse gas emission abatement potential and associated costs of integrating renewable and low carbon energy technologies into the Canadian Oil Sands," J. Clean. Prod., vol. 272, p. 122820, 2020, https://doi.org/10.1016/j.jclepro.2020.122820.
47. M. Zafar, A. Khan, and H. A. Khan, "Harvesting Energy from Ocean: Technologies and Perspectives," Energies, vol. 15, p. 3456, 2022, https://doi.org/10.3390/en15093456.
48. X. Zhang, L. Zhang, Y. Yuan, and Q. Zhai, "Life Cycle Assessment on Wave and Tidal Energy Systems: A Review of Current Methodological Practice," Int. J. Environ. Res. Public Health, vol. 17, 2020, https://doi.org/10.3390/ijerph17051604.
49. N. Rambabu and V. K. Srineash, "A Review on Methodologies to Upgrade the Coastal Structures to Enhance the Coastal Resilience," Environ. Sci. Eng., 2022, https://doi.org/10.1109/OCEANSChennai45887.2022.9775524.
50. M. A. Lacasse, A. Gaur, and T. V Moore, "Durability and Climate Change — Implications for Service Life Prediction and the Maintainability of Buildings," Buildings, 2020, https://doi.org/10.3390/buildings10030053.
51. K. Henna, A. Saifudeen, and M. Mani, "Resilience of vernacular and modernising dwellings in three climatic zones to climate change," Sci. Rep., no. 0123456789, pp. 1–14, 2021, https://doi.org/10.1038/s41598-021-87772-0.
52. A. E. Ioannou and C. S. Laspidou, "Resilience Analysis Framework for a Water—Energy—Food Nexus System Under Climate Change," Front. Environ. Sci., no. May, pp. 1–14, 2022, https://doi.org/10.3389/fenvs.2022.820125.
53. M. Chen and K. Caldeira, "Climate change as an incentive for future human migration," Earth Syst. Dyn., pp. 875–883, 2020, https://doi.org/10.5194/esd-11-876-2020.
54. H. Kirk, J. Di Francesco, D. Johnstone, L. Yuri, D. Marco, and W. Duan, "Migration towards Bangladesh coastlines projected to increase with sea-level rise through OPEN ACCESS Migration towards Bangladesh coastlines projected to increase with sea-level rise through 2100," Environemental Resaerch Lett., 2021, https://doi.org/10.1088/1748-9326/abdc5b.
55. L. De Mello and J. Martinez-Vazquez, "Climate Change Implications for the Public Finances and Fiscal Policy: An Agenda for Future Research and Filling the Gaps in Scholarly Work," Economics, pp. 194–198, 2022, https://doi.org/10.1515/econ-2022-0026.
56. J. Bjordal, T. Storelvmo, and A. A. S. Jr, "Quantifying uncertainty about global and regional economic impacts of climate change OPEN ACCESS Quantifying uncertainty about global and regional economic impacts of climate change," Environ. Res. Lett., vol. 17, 2022, https://doi.org/10.1088/1748-9326/ac8ab1.
57. M. Andersson and C. Baccianti, "Occasional Paper Series Climate change and the macro economy." European Central Bank (ECB), 2020.
58. R. Wolniak and B. Skotnicka-Zasadzie, "Development of Photovoltaic Energy in EU Countries as an," Energies, vol. 15, p. 662, 2022, https://doi.org/10.3390/en15020662.
59. W. Buchholz, L. Dippl, and M. Eichenseer, "Subsidizing renewables as part of taking leadership in international climate policy: The German case," Energy Policy, vol. 129, no. February, pp. 765–773, 2019, https://doi.org/10.1016/j.enpol.2019.02.044.
60. F. Harke and P. Otto, "Solar Self-Sufficient Households as a Driving Factor for Sustainability Transformation," Sustainability, 2023, 10.3390/ su15032734 Academic.
61. K. Johansen, "Wind energy in Denmark," IEEE power & energy magazine, no. june, pp. 94–102, May 2021.

62. M. C. Mendoza, E. Martinez-Gomez, and H. Li, "Feasibility Analysis of Promoting Wind and Solar Energy in Texas Farms," Sustainability, 2022, https://doi.org/10.3390/su14042222.
63. L. Welbanks et al., "The Effects of Wind Veer During the Morning and Evening Transitions the Effects of Wind Veer During the Morning and Evening Transitions," J. Phys. Conf. Ser., pp. 0–15, 2019, https://doi.org/10.1088/1742-6596/1452/1/012075.
64. A. S. da Silva, I. Vera, L. L. Soares, and A. Maria Eugenia, "Case Study: Itaipu and SDG 17," 2019.
65. H. Kongsjorden, O. Kdrstad, and T. A. Torp, "Saline Aquifer Storage of Carbon Dioxide in the Sleipner Project," Waste Manag., vol. 17, no. 5, pp. 303–308, 1998.
66. S. Gardarsdottir, K. Jordal, K. Lundqvist, and M. Biermann, "Preem CCS—a pioneering Swedish-Norwegian collaboration showcasing the full CCS chain," in 15th International Conference on Greenhouse Gas Control Technologies, GHGT-15 1, 2021, no. March, pp. 1–6.
67. M. Biermann, S. Harvey, J. Kjärstad, F. Johnsson, A. Reyes-lúa, and O. Stefania, "Lessons learned from the Preem-CCS project—a pioneering Swedish-Norwegian collaboration showcasing the full CCS chain," in 16th International Conference on Greenhouse Gas Control Technologies, GHGT-16, 2022, no. October.
68. O. Miyamoto, C. Maas, T. Tsujiuchi, M. Inui, and T. Hirata, "KM CDR Process TM Project Update and the New Novel Solvent Development," Energy Procedia, vol. 114, no. November 2016, pp. 5616–5623, 2017, https://doi.org/10.1016/j.egypro.2017.03.1700.
69. A. Manuilova, J. Koiwanit, L. Piewkhaow, M. Wilson, C. W. Chan, and P. Tontiwachwuthikul, "Life Cycle Assessment of Post-Combustion CO_2 Capture and CO_2 - Enhanced Oil Recovery based on the Boundary Dam Integrated Carbon Capture and Storage Demonstration Project in Saskatchewan," Energy Procedia, vol. 63, pp. 7398–7407, 2014, https://doi.org/10.1016/j.egypro.2014.11.776.
70. B. H. F. Misch, Y. Camara, "Electric Vehicles, Tax incentives and Emissions: Evidence from Norway," 2021. https://doi.org/10.5089/9781513582818.001.
71. C. Mouli et al., "Economic and CO_2 Emission Benefits of a Solar Powered Electric Vehicle Charging Station for Workplaces in the Netherlands," IEEE Transp. Electrif. Conf. Expo, 2016, https://doi.org/10.1109/ITEC.2016.7520273.
72. L. Quitzow, W. Canzler, P. Grundmann, M. Leibenath, T. Moss, and T. Rave, "The German Energiewende e What' s Happening? Introducing the Special Issue," Util. Policy, vol. 41, pp. 163–171, 2016, https://doi.org/10.1016/j.jup.2016.03.002.
73. S. Ashina and T. Nakata, "Energy-efficiency strategy for CO_2 emissions in a residential sector in Japan," Appl. Energy, vol. 85, pp. 101–114, 2008, https://doi.org/10.1016/j.apenergy.2007.06.011.
74. M. C. Sanchez, R. E. Brown, C. Webber, and G. K. Homan, "Savings estimates for the United States Environmental Protection Agency' s ENERGY STAR voluntary product labeling program," Energy Policy, vol. 36, pp. 2098–2108, 2008, https://doi.org/10.1016/j.enpol.2008.02.021.
75. A. Patel, "Climate change & agriculture in India- effective implementation of national mission for sustainable agriculture," Int. J. Res. -GRANTHAALAYAH, vol. 4, no. 11, pp. 52–71, 2016, https://doi.org/10.29121/granthaalayah.v4.i11.2016.2420.
76. S. Wales et al., "The strategic use of minimum tillage within conservation agriculture in southern New South Wales, Australia," Soil Tillage Res., vol. 193, no. March, pp. 17–26, 2019, https://doi.org/10.1016/j.still.2019.05.021.
77. R. Bramley and S. A. M. Trengove, "Precision agriculture in Australia: present status and recent developments," Eng. Agric., pp. 575–588, 2013, https://doi.org/10.1590/S0100-69162013000300014.
78. J. C. Polidoro et al., "Potential impact of plans and policies based on the principles of conservation agriculture on the control of soil erosion in Brazil," L. Degrad. Dev., vol. 32, no. February, pp. 1–12, 2021, https://doi.org/10.1002/ldr.3876.

Energy Statistics Influence on Energy Consumption for Oil-Rich Countries

Fotouh Al-Ragom

Abstract This chapter thoroughly explores the intricate relationship between energy, economics, and geopolitics in oil-rich countries, highlighting the complex dynamics of energy consumption and its global significance. Through a comparative analysis of energy intensities among nations, it emphasizes the crucial role of the energy mix in shaping consumption patterns, with technology playing a pivotal role in resource management. The discussion extends to the policy landscape, investigating how governmental decisions and regulations shape energy trajectories, emphasizing the crucial role of national policies in promoting sustainability, fostering innovation, and addressing environmental concerns. The exploration broadens to consider key factors contributing to the global influence of oil-rich countries, examining aspects such as energy supply, global oil prices, and the economic and fiscal impact, including foreign exchange reserves and geopolitical influence. It further encompasses the realm of energy transitions, investigating the influence of infrastructure investments, global trade dynamics, and diplomatic engagements, while also examining energy policy shifts and shedding light on the intricacies of energy intensity and the evolving energy mix.

Moreover, the chapter presents a detailed analysis of energy transitions, emphasizing the pivotal role of technology innovation. It explores key factors influencing energy intensity, considering the nuanced interplay of technological advancements, policy decisions, and economic imperatives. In delving into these interconnected dimensions, this chapter provides a comprehensive understanding of the global influence wielded by oil-rich countries and their indispensable role in shaping the future of energy on the international stage.

Keywords Energy mix · Economic growth dependence · Energy intensity comparison · Technological factors · Geo-political influence

F. Al-Ragom (✉)
Energy and Building Research Center, Kuwait Institute for Scientific Research, Kuwait city, Kuwait
e-mail: fragom@kisr.edu.kw

K. J. Sreekanth (ed.), *Energy Efficiency Improvements with Emission Abatement for Energy Sustainability*, https://doi.org/10.1007/978-981-97-9480-5_5

Introduction

Oil-rich countries, often referred to as petroleum-exporting nations, have a considerable impact on the global energy landscape as they exhibit common energy trends that shape their economic landscapes. These nations heavily rely on fossil fuels, particularly oil and natural gas, for both domestic consumption and export. Energy production is often export-oriented, generating substantial revenues. The economic growth of these nations is intricately linked to global oil prices and export volumes, resulting in varying energy consumption patterns. Some oil-rich countries provide energy subsidies to citizens, fostering artificially low prices and high consumption rates. The focus on energy-intensive industries and lifestyles contributes to high energy intensity. Limited diversification of the energy mix leaves these nations vulnerable to supply disruptions and price fluctuations. Environmental challenges emerge from high emissions and dependence on non-renewable sources. Developed oil-rich nations invest in technological advancements, with a focus on fossil fuel extraction and renewable energy development.

The relationship between energy statistics, energy consumption, and oil-rich economies is dynamic. This relationship is influenced by several factors including energy mix, economic growth, energy intensity, and technological advancements. This chapter delves into these crucial aspects, providing a comprehensive understanding of how energy statistics shape energy consumption trends in oil-rich countries.

Oil-Rich Countries Statistics

Nations with significant reserves of crude oil are referred to as oil-rich countries or petro states. These countries usually have a substantial portion of their government revenue and exports derived from oil production and exports. They are also characterized by their strong dependence on oil-related activities and are typically major players in the global oil market. The wealth generated from oil resources can significantly impact these nations' economies and geopolitical influence, earnings making them key players in the energy industry and international relations. Among these countries are Saudi Arabia, Russia, Iraq, Iran, Kuwait, the United Arab Emirates, Venezuela, Libya, Nigeria, and Norway. The United States and Canada can also be considered among these nations due to their shale oil and oil sands resources. The top ten countries in terms of confirmed oil reserves as of 2022 (shown in Fig. 1) are vital to both economic stability and global energy security because they have sizable reserves that affect global energy markets. In addition, these reserves play an important role in shaping the geopolitical landscape and energy policies, as they contribute to meeting the world's growing energy demands.

In terms of oil output, these reserves have a significant impact on pricing and supply trends. Leading countries' global oil production is shown in Fig. 2. The graph

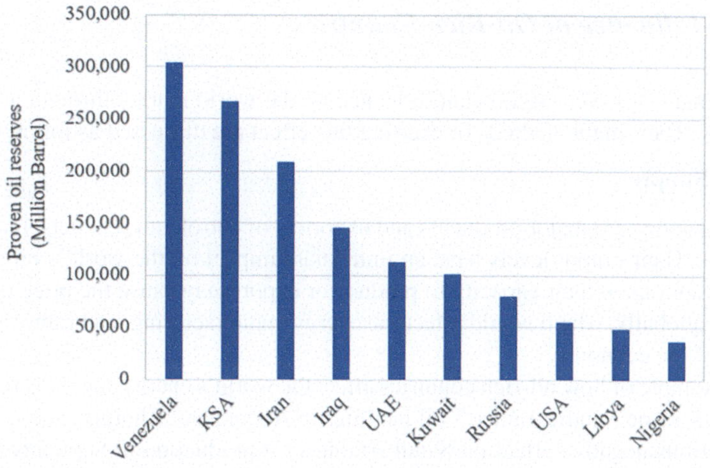

Fig. 1 Proven crude oil reserves for the top ten countries in 2022 [1]

shows that in 2022, the United States accounted for 16% of global oil production, making it the leading oil producer in the world, followed by Saudi Arabia at 14.6%.

The United States (USA) high share of shale oil production and the repeal of the crude oil export ban in 2015 that allowed USA crude oil to be freely marketed and sold on the global market contributed to the USA exports increase [2]. The significant portion of shale oil production in the United States, coupled with the lifting of the crude oil export ban in 2015, allowed for the free global marketing and sale of USA crude oil, which in turn supported the USA export surge. In the first half of 2023, USA crude exports achieved a historic peak, averaging 3.99 million barrels per day (b/d), representing the highest level since 2015 [3]. This places the USA among the top crude oil-producing countries.

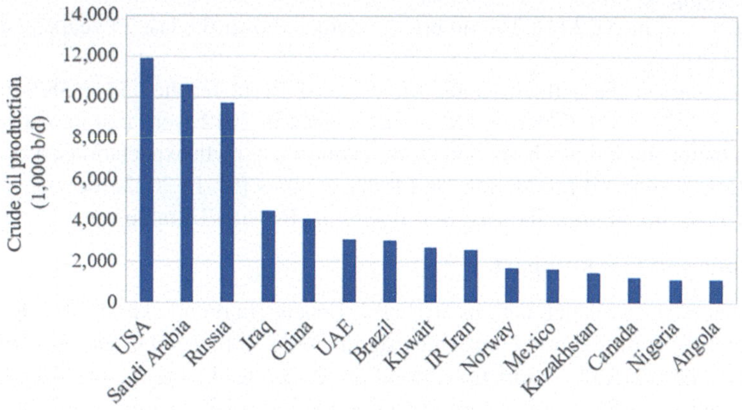

Fig. 2 World crude oil production by country in 2022 [1]

Global Influence of Oil-Rich Countries

Oil-rich nations have a significant influence on the world market through a variety of means. Their main methods of exerting this effect are discussed as follows.

Energy Supply

Oil-rich nations are major producers and exporters of petroleum products, including crude oil. Their output levels have an immediate impact on the world's oil supply. Any disruptions in their capacity to produce or export may cause the price of oil to fluctuate globally, which would affect the energy, manufacturing, and transportation sectors of the economy.

An example of how oil-rich countries affect the world's energy supply is the 2019 drone attacks on Saudi Aramco's oil facilities in Abqaiq and Khurais. These attacks had a serious negative effect on Saudi Arabia's oil production in September 2019, forcing a temporary stop to a large amount of the nation's oil output. Saudi Arabia is a significant exporter and one of the world's top producers of oil. The attack disrupted a substantial 5.7 million barrels per day (Mb/d) of Saudi oil production and 0.7 Mb/d of natural gas liquids production, marking the most significant outage in terms of volume in the modern history of oil. This attack's immediate effect was to disrupt the world's oil supply, which caused oil prices to soar on international markets from $60 to $69 per barrel [4].

Global Oil Prices

Petro states frequently work together to regulate oil production and maintain oil prices, particularly those that are members of the Organization of the Petroleum Exporting Countries (OPEC). Their choices regarding oil production quotas can have a significant effect on oil prices globally, influencing consumer spending, energy prices, and inflation rates. OPEC sets the oil production allocations for its members. Table 1 lists some allocations for top OPEC-producing countries [1].

According to OPEC, the nominal oil process increased from $69.98/barrel to $100.08/barrel in 2022 [1]. The oil prices fluctuated over the last 25 years as shown in Fig. 3 [5].

Reflecting on these fluctuations, the price of Brent reached $111.58/barrel in 2012 and 2022 $100.93/barrel. Oil price changes in 2012 were mainly caused by various factors including changes in global economic growth expectations, oil supply disruptions, rising oil production, and Iran sanctions [6]. In 2022, the main factor that affected the oil price increase was due to the Russian/Ukraine war.

Economic and Fiscal Impact

The economies of oil-rich nations are highly dependent on oil exports. The revenue generated from oil exports contributes significantly to their GDP and government budgets. Consequently, fluctuations in oil prices can lead to economic volatility in these countries, which can, in turn, affect global financial markets. The economies of oil-rich nations are tied to the fortunes of their oil exports. The revenue generated

Table 1 OPEC Members' crude oil production allocations (1,000 b/d)

Country	Allocations (1,000 b/d)
OPEC	25,416
Saudi Arabia	10,478
Iraq	4,431
United Arab Emirates	3,019
Kuwait	2,676
Nigeria	1,742
Angola	1,455
Algeria	1,007
Congo	310
Gabon	177
Equatorial Guinea	121

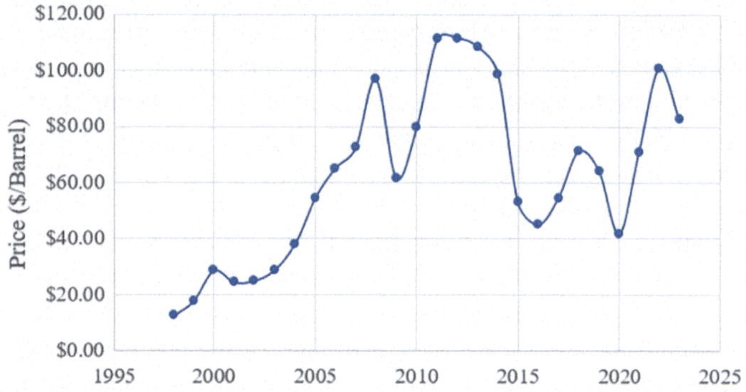

Fig. 3 Brent crude oil prices [5].

from the export of crude oil plays a pivotal role in these nations, often contributing a substantial portion to their Gross Domestic Product (GDP) and government budgets. Rising oil prices can be advantageous for oil-producing nations, supporting their economic stability and allowing their governments to make investments in a range of industries and public services. As an example, Saudi Arabia one of the world's largest oil producers, benefited from rising oil prices. Higher oil revenues have allowed the Saudi government to invest in infrastructure projects, diversify its economy through the Vision 2030 plan, and improve public services such as healthcare, education, and transportation. Saudi initiatives focusing on tourism, small and medium-sized enterprises, and "Saudi Made" emerged and contributed to the national GDP [7]. Norway capitalized on its oil revenue and managed its wealth primarily through the Government Pension Fund Global (GPFG). The income revenue gained from investing in oil revenues has been used to invest in infrastructure, education, and healthcare, ensuring economic stability and high quality of life for its population [8]. Similarly, in Kuwait, oil revenues have been contributing to savings for future generations, developing infrastructure, and diversifying its economy through initiatives like the Kuwait Vision 2035 plan [9]. These examples show how countries with an abundance of oil resources can use the spikes in oil prices to strengthen public services, accelerate economic growth, and direct investments into various industries. A decline in oil prices can have a significant effect on the economy and finances. Decreased oil revenue can put pressure on government budgets, resulting in cuts to public spending, austerity measures, and even deficits. This can be especially difficult in nations where oil revenue makes up a sizable portion of the national budget.

An example is what happened in Venezuela in mid-2010. Since the Venezuelan economy heavily relies on oil exports, and when oil prices plummeted in the mid-2010s, the country's revenue significantly declined. An economic crisis, severe budget deficits, and hyperinflation were the results of this significant fiscal and economic impact. Widespread social and political unrest resulted from the government's forced cost-cutting measures and decreased ability to fund public services due to lower oil revenues. Venezuela provides an example of the negative impact of falling oil prices on an oil-rich country. The Venezuelan economy heavily relies on oil exports, and when oil prices plummeted in the mid-2010s, the country's revenue significantly declined. This had a profound economic and fiscal impact, leading to severe budget deficits, hyperinflation, and an economic crisis. Diminished oil income placed pressure on the government's capacity to support public services, resulting in extreme actions that triggered extensive social and political unrest [10].

These above-mentioned examples highlight the contrasting economic outcomes in oil-rich countries when oil prices rise and fall. While rising prices can lead to prosperity and economic stability, falling prices can expose vulnerabilities and necessitate fiscal adjustments to mitigate the impact on government budgets and public services. Thus, it's essential for these nations to also plan for the potential volatility of oil markets and consider strategies for long-term economic sustainability beyond oil. some oil-producing nations have sought to diversify their economies, reducing their dependence on oil exports and mitigating the impact of oil price fluctuations.

The economic and fiscal fluctuations have a ripple effect and can extend beyond the borders of oil-rich nations. Global financial markets are sensitive to oil price movements and can trigger uncertainty and volatility in financial markets, affecting the valuations of energy companies, impacting stock indices, and even influencing currency exchange rates.

Foreign Exchange Reserves

Nations with a high reliance on oil exports gain significant foreign exchange reserves in addition to receiving essential income from sales transactions. The size of these reserves is crucial in determining how national currencies are valued and how international capital flows are shaped. These reserves are frequently dispersed carefully, with a sizeable amount invested in different global financial markets to yield returns and diversify the country's asset holdings. Several oil-rich countries established beneficiary funds for the country's nationals that are financed by the country's oil revenues. As an example, Norway established a Government Pension Fund Global (GPFG). The fund is one of the largest sovereign wealth funds globally. It invests in a diverse range of assets, including stocks, bonds, and real estate, contributing to Norway's economic stability [11]. The United Arab Emirates (UAE), particularly Abu Dhabi, has leveraged its oil wealth to establish sovereign wealth funds like the Abu Dhabi Investment Authority (ADIA). These funds actively invest in global markets, contributing to the country's economic resilience and providing a buffer against oil market uncertainties [12].

Geo-Political Influence

The geopolitical stability of oil-rich regions is of great concern to the global community. Disputes, conflicts, or changes in leadership in these areas can disrupt oil supplies and cause uncertainty in energy markets.

Nerlinger and Utz examined the impact of the war between Russia and Ukraine on the stock prices of energy companies, looking at a global sample of 1630 energy companies from 75 countries. They discovered that in 2022, the energy sector firms did better than the stock market. The American companies outperformed the European and Asian ones [13]. The substantial change in energy commodity prices, shown in Fig. 4, asserts the geopolitical stability influence. After the Russian invasion, oil and gas prices experienced a modest increase, while coal prices surged by 150%.

They have opined that given how the war affected companies in the renewable energy sector, these companies underperformed fossil and nuclear energy after a brief period of growth a few days following the invasion. These findings might also add to the discussion of Hook and Hume [14] in the Financial Times on whether the Ukraine war will derail the green energy transition. Investors seem to still believe in

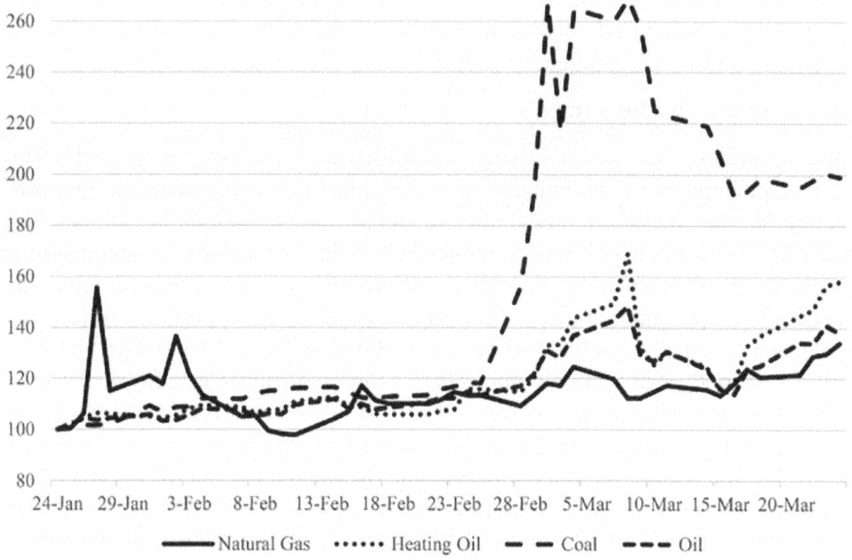

Fig. 4 Energy prices around the Russia and Ukraine war [13]

the business concepts of fossil fuel and uranium-based energy technology companies, as seen by their outperformance on the stock market.

Energy Transitions

Oil-rich countries are increasingly diversifying their economies and energy sources due to concerns about the long-term viability of fossil fuels and the growing global focus on sustainability. Their investments in renewable energy and technologies also impact the global energy market.

Hancock expressed her view in the Financial Times, given the Ukraine war and Biden's IRA forcing the European Union (EU) to accelerate the energy transition [15]. The EU's dependency and reliance on Russian fossil fuels was the main driver in accelerating the energy transition scenarios after the Russian-Ukraine war. In the State of the Energy Union Report 2023, the European Union (EU) explains how it accelerated the transition to clean energy, diversified its supply, and conserved energy in response to Russia's aggression in Ukraine and weaponization of its energy supplies. In May 2022, the European Commission significantly increased its targets for clean energy capacity. By 2030, legislative targets were set to guarantee that the European Union will have at least 42.5% renewable energy by 2030, with a higher objective of 45%. Targets for energy efficiency were also increased, to reduce total energy use by 11.7% by 2030 [16, 17].

In 2021, Denmark, a member state of the European Union, produced a total of 3.8 million cubic meters of crude oil. Its crude oil production declined from 137,000 b/d to 66,000 b/d from 2017 to 2021 respectively [1]. According to IEA, the Danish government decided to phase out their fossil fuel extraction by 2050 [18].

In Saudi Arabia, guided by Vision 2030 [19], and national reforms, national reforms were introduced to decrease energy consumption, introduce renewable energy, and lower energy intensity [20].

Infrastructure and Investments

These countries often invest in energy infrastructure and projects globally. This can include financing or building oil refineries, pipelines, and transportation infrastructure in other nations, thereby shaping global energy logistics. As an example, Saudi Arabia invests in building oil refineries in Asia. According to Reuters, India is seeking the involvement of Saudi Arabia's Aramco in its proposed strategic petroleum reserve (SPR) program, aiming to enhance relations with its primary oil provider [21]. Qatar invests in liquefied natural gas (LNG) globally mainly through QatarEnergy. QatarEnergy has acquired interests in a variety of upstream assets in several nations throughout the world over time. Most of these are now working on exploratory projects in offshore blocks, and some are working with operational assets. Argentina, Brazil, Mexico, Congo, Cyprus, Mozambique, Kenya, Oman, and Morocco [22].

Kuwait invests in oil exploration in various countries through the Kuwait Foreign Petroleum Exploration Company (KUFPEC), a subsidiary of Kuwait Petroleum Corporation (KPC) [23]. Within the Asia and Euro-Africa Region of KUFPEC, there are projects in Pakistan, Malaysia, Indonesia, Norway, Egypt, and Yemen. KUFPEC has several projects in both Australia and Canada in the Canada-Australia Region

[24]. These examples showcase the diverse strategies adopted by petro states to secure their interests and influence the global energy landscape through infrastructure investments.

Global Trade and Diplomacy

The energy trade from oil-rich nations can impact international trade relations and diplomacy. Bilateral and multilateral agreements concerning oil trade have geopolitical implications.

Russian gas and oil imports are an important source of energy for the European Union (EU). Diplomatic relations are impacted by energy deals between Russia and the EU, which have geopolitical implications. It has occasionally strained relations and impacted diplomatic decisions to have disagreements about energy costs or supply disruptions. The Russia-Ukraine is an example of how global trade and diplomacy can be affected. Another example is the Organization of the Petroleum Exporting Countries (OPEC) and OPEC+. OPEC+ is an alliance between OPEC members and several non-OPEC oil-producing nations. These non-OPEC members take part in the organization's programs, like voluntary supply reductions. Among the OPEC+ members are Russia, Mexico, and Kazakhstan. OPEC nations, including major oil producers like Saudi Arabia and Russia, engage in discussions and agreements to manage oil production levels. These agreements have geopolitical implications as they influence global oil prices, impacting both oil-importing and oil-exporting nations [25, 26].

In summary, the influence of oil-rich countries on the global market is substantial, not only because of their role in energy production and supply but also due to the economic, political, and financial consequences of their activities. Changes and developments in these nations have far-reaching effects on the stability and functioning of the global market.

Energy Policy Shifts

Some oil-rich nations are gradually shifting their energy policies to reduce dependence on fossil fuels and promote sustainability. As an example, Norway as an oil producer, implemented policies to promote sustainable transportation. The country started incentivizing electric vehicles (EVs) as early as 1990 by offering incentives, tax breaks, and exemptions [27]. The incentives included reduced annual road tax and exemption from purchase tax. Free car parking was also introduced as well as free access to bus lanes and the roads toll exemption. Accordingly, Norway emerged as the world's top-ranked country in 2021 with a 64.5% market share of BEVs, up from 12.5% in 2014 [28, 29].

Energy Intensity

Energy intensity is an indicator that measures the efficiency of energy use in an economy or a specific sector [30]. It is typically expressed as the amount of energy consumed per unit of economic output, often measured in terms of gross domestic product (GDP). The purpose of energy intensity is to assess how much energy is required to generate a certain level of economic activity and compare different countries. By comparing energy intensity values between nations, policymakers, researchers, and analysts can gain insights into the relative energy efficiency of economies, identify best practices, and formulate strategies to improve overall energy performance on a global scale. This comparative analysis helps in evaluating the impact of energy policies, technological innovations, and industrial practices across countries, facilitating informed decision-making to enhance energy sustainability and reduce environmental impacts worldwide. The intensity is defined as follows:

$$Energy\,Intensity = \frac{Total\,Energy\,Consumption}{Economic\,Output\,(Ex.GDP)}$$

A lower energy intensity value denotes a more energy-efficient use of resources, implying that the sector or economy is generating more output for each energy unit used. A higher energy intensity value, on the other hand, may indicate poorer efficiency and thus more energy consumption for a given level of economic activity. According to the World Bank, the energy intensity per Purchasing Power Parity Gross Domestic Product (PPP DGP) of Oil-rich (OPEC and OPEC+) countries is shown in Fig. 5. Bahrain ranked the highest followed by Iran then Kuwait with an energy intensity level of primary energy of 9.47, 8.78, and 8.47 MJ/$2017 PPP GDP respectively. The high energy intensity of these countries can be attributed to the availability of electricity subsidies, limited awareness regarding energy efficiency, and the absence of comprehensive energy policies and incentive schemes. As an example, according to the Head of the National Productivity Organization of Iran, Mir-Saman Pishvaei, "Iran had the world's biggest per capita energy consumption in 2022". He expressed his concern that Iran's consumption is trending upward, while the world's energy consumption is trending downward. Pishvaei's urging of Iranian officials to take serious measures in energy consumption highlights the evident lack of energy efficiency within the country [31]. On the other hand, Angola, which has the lowest energy intensity, had a low GDP per capita in 2022 of about $3000 as indicated by the World Bank [32]. The economy of Angola ranked as the third-largest in sub-Saharan Africa, is predominantly shaped by the oil and gas sector industry. This industry constitutes approximately 50% of the country's GDP and stands as the principal revenue source, contributing to over 70% of government revenue and constituting 90% of Angola's total exports. Yet its electricity consumption is considerably small as in 2021, electricity consumption was 476 kWh/person.

Comparing that to Kuwait, the electricity consumption in the same year was about 15,815 kWh/person [33]. Even though Kuwait enforces an energy conservation code of practice in buildings, Kuwait's electricity is highly subsidized for all sectors by the

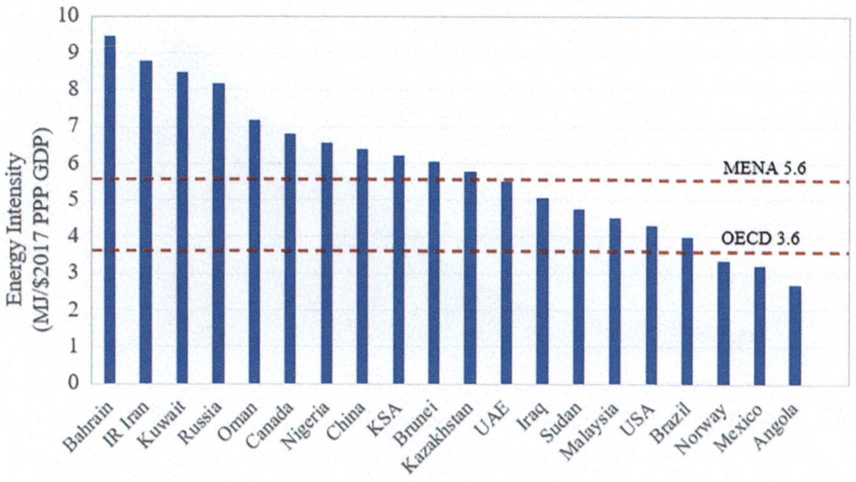

Fig. 5 Energy intensity level of primary energy in 2020 [35]

government. The energy intensity of the Organization for Economic Co-operation and Development (OCED) was 3.6 MJ/$2017 PPP GDP in 2020. The OECD is an international organization with 38 members that originated in 1961 [34]. The majority of the OECD members (82%) are developed countries. In most of the selected countries, energy intensity was higher than that of the average OECD. The USA and Canada have energy intensities that are higher than the OECD average, even though they are both OECD members. In view of the Middle East and North Africa (MENA) countries' 2020 energy intensity of 5.8 MJ/$2017 PPP GDP, the selected countries mainly require higher energy intensities except UAE, Iraq, and Sudan.

Energy Mix

The term "energy mix" refers to the combination of different energy sources that a region, country, or entity uses to meet its energy needs. This mix typically includes various forms of energy, such as fossil fuels (coal, oil, natural gas), renewable energy sources (solar, wind, hydroelectric, geothermal), and nuclear power.

The geographic location, the availability of natural resources, the economy, and governmental regulations can all have a significant impact on the energy mix composition in a given location. Because it can improve energy security, lessen its negative effects on the environment, and provide a more resilient and sustainable energy infrastructure, a diversified energy mix is frequently seen as desirable.

Given the world's energy mix, shown in Fig. 6, it's evident that the dominant source is oil, followed by coal then gas.

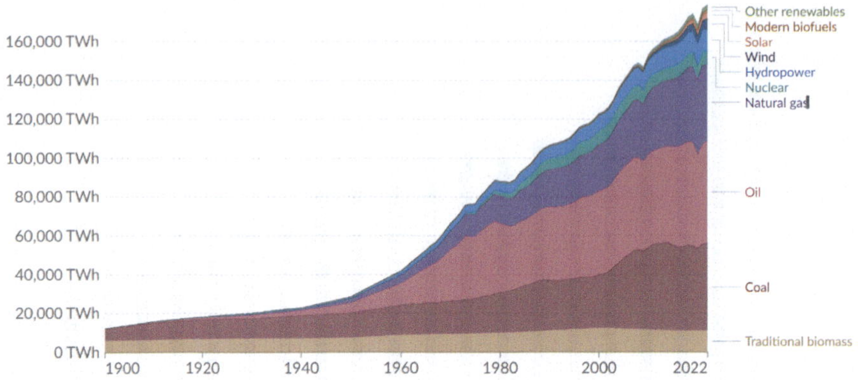

Fig. 6 Global primary energy consumption by source [36]

As shown in Table 2, oil and gas are the dominant energy source in most of the selected OPEC and OPEC+ countries. Coal is the major energy source in China and Kazakhstan, while Norway is mainly dependent on hydropower. Iran, Malaysia, Oman, Russia, and the UAE primarily rely on natural gas as their main energy source. When considering the overall energy consumption in 2022, China claimed the top position, recording 44,275.91 terawatt-hours (TWh), followed by the USA and Russia with 26,641.78 TWh and 8,025.71 TWh, respectively. Kuwait is mainly dependent on oil and gas with a minimal contribution from renewable energy. In fact, according to Kuwait's Energy Outlook, Kuwait is using solar energy with a share just short of 1% of the total energy mix [22]. The top country in terms of the use of solar is China, with 2.52% of the total energy mix. In fact, China had 393.03 GW of installed solar capacity while the world total was 1053.12 GW in 2022 [36].

The average solar energy was about 1 TWh. Norway leads in hydropower energy, with Brazil and Canada following suit. The foremost utilization of wind energy is observed in Norway, succeeded by Brazil and China. Brazil stands out for the highest application of biofuels. In terms of nuclear energy usage, the USA takes the lead, trailed by Russia, Canada, UAE, China, Mexico, and Iran.

Among the developing nations are the USA, Canada, and Norway, each boasting a diversified energy mix. Notably, Norway heavily relies on hydropower and does not incorporate nuclear energy. The USA and Canada maintain a relatively equal distribution of oil and gas in their energy portfolios. Gulf Cooperation Council (GCC) Countries, including Kuwait, Oman, UAE, and Saudi Arabia have a low share of clean energy in their energy mix.

As a developing nation, Mexico draws from various energy sources, with oil and gas prominently featured. China, while predominantly reliant on coal, maintains a diverse energy portfolio encompassing oil, gas, nuclear, and renewables. Regarding renewable energy, hydropower is predominantly utilized in the selected countries except those that experience water scarcity.

Table 2 Energy mix of OPEC and OPEX+ selected countries in 2022 [36]

Country	Energy source (%)								
	Other renewables	Biofuels	Solar	Wind	Hydropower	Nuclear	Gas	Coal	Oil
Brazil	4.52	6.5	2.11	5.71	29.89	0.98	8.58	4.37	*37.34*
Canada	0.69	0.6	0.4	2.49	26.44	5.52	*30.96*	2.73	30.17
China	1.27	0.06	2.52	4.49	7.67	2.36	8.49	*55.47*	17.67
IR Iran	0	0	0.06	0.1	0.58	0.49	*67.79*	0.63	30.37
Iraq	0	0	0.17	0	1.34	0	29.46	0	*69.03*
Kazakhstan	0	0	0.51	0.76	2.77	0	25.04	*45.98*	24.94
Kuwait	0	0.12	0	0	0	0	49.13	0.28	*50.47*
Malaysia	0.27	1.04	0.53	0	6.31	0	*36.76*	19.39	35.7
Mexico	0.74	0.11	2.07	2.19	3.84	1.12	39.85	2.88	*47.2*
Norway	0.16	0.83	0.13	7.34	*63.16*	0	7.57	1.77	19.03
Oman	0	0	0.92	0.07	0	0	*68.66*	0.24	30.11
Russia	0.03	0.05	0.08	0.14	6.42	6.97	*50.84*	11.06	24.42
Saudi Arabia	0	0	0.07	0	0	0	37.69	0.04	*62.2*
UAE	0	0	1.29	0	0	3.59	*49.79*	2.03	43.29
USA	0.84	1.63	2.02	4.3	2.53	7.63	33.08	10.29	*37.69*

Note The main energy source is highlighted

Looking at the electricity consumption in some selected Oil-Rich countries, as shown in Fig. 7. Norway consumed about 28 MWh/person in 2022 followed by Bahrain then Kuwait with 21 MWh/person and 17 MWh/person respectively. Norway's per capita energy consumption trend is fluctuating, the USA and Canada is trending downward while China, Bahrain, Kuwait UAE, and Saudi Arabia are trending upward.

Energy consumption can be affected by various factors including population growth, climate and weather conditions, government policies, technological advancements, urbanization, economic growth, and consumer behavior and awareness. Keho studied the drivers of energy consumption in developing countries and concluded that in some selected South African countries, economic growth, industrial output, and population increase energy consumption [37]. Komarova et al. examined the energy consumption levels of different nations and their relationship to economic progress [38]. It was determined that non-OECD countries' GDP is more impacted by energy usage because of the extensive nature of economic development. They listed the top four countries in terms of 2019 per capita energy consumption as Qatar, Singapore, Iceland, and the UAE with 16.8 toe/capita, 14.8 toe/capita, 14.3 toe/capita, 11.2 toe/capita) respectively. They also shared that there is significant energy consumption in Saudi Arabia, the USA, Kuwait, Canada, and Norway which was 7.5 toe/capita,

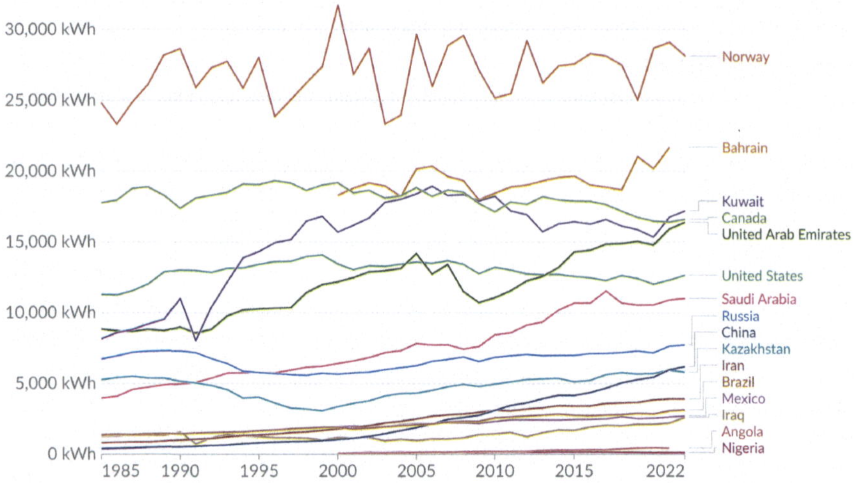

Fig. 7 Annual average per capita electricity generation [36]

6.9 toe/capita, 9.9 toe/capita, 8.9 toe/capita, 7.8 toe/capita, and 6.7 toe/capita respectively. They explained that such a consumption trend is because these countries are characterized by having vast reserves and supplies of fossil fuels of their own. To guarantee domestic consumption, they also engage in the large-scale extraction of primary energy sources which exacerbates their energy consumption. Conversely, in nations like Singapore, South Korea, and Belgium, where the population is small, the high level of energy consumption can be attributed to the growth of institutional factors and international trade.

Energy Transition

Challenges that governments and societies must overcome to achieve a secure and sustainable energy system are linked. The term "energy trilemma" refers to these difficulties. The World Energy Council is credited with coining the phrase [39, 40]. These three challenges include energy security, energy equity, and environmental sustainability. These three challenges are explained as follows:

1. Energy Security: Ensuring a dependable and secure energy supply to fulfill current and future demands is known as energy security. This entails increasing the resilience of energy infrastructure, diversifying energy sources, and reducing the risk of supply disruptions.
2. Energy Equity: Energy equity or accessibility refers to giving everyone access to reasonably priced energy, solving the problem of energy poverty, and making sure that energy resources are allocated equitably among various socioeconomic groups and geographical areas.

3. Environmental Sustainability: To reduce harmful effects on the environment and slow down climate change, ecologically sustainable methods of energy generation and consumption should be encouraged. This entails boosting renewable energy sources, cutting greenhouse gas emissions, and enhancing energy efficiency.

The energy trilemma highlights the interconnectedness and sometimes contradictory nature of these three goals. The trade-offs and opportunities for cooperation between energy security, equity, and sustainability must be carefully considered to arrive at a balanced solution. Advancement in one field could lead to challenges in another. Stakeholders such as governments, corporations, and communities must work together to develop policies and programs that address these concerns holistically. Discussions concerning global energy policy and the transition to a more resilient and sustainable energy future are common.

Energy transitions are predicted to be an all-encompassing strategy that can resolve the energy trilemma. Transitioning to more sustainable and renewable energy sources from conventional fossil fuel-based ones is necessary for the global energy system. To lessen the impact on the environment, provide energy security, and foster economic sustainability, it entails a fundamental shift in the ways that energy is produced, delivered, and used.

A roadmap for transforming the energy sector to meet the goals of the Paris Agreement is outlined in the World Energy Transitions Outlook. By the middle of the century, it provides a route that aims to achieve net-zero carbon dioxide emissions and limit the increase in global temperature to 1.5 °C [41]. To meet the 1.5 °C target, the proportion of renewables in electricity generation should rise from the present 28 to 68% by 2030 and 91% by 2050. The rate of improvement in energy intensity should reach 3.3% by 2030 and 2.8% by 2050, compared to the current 1.7%. Additionally, clean hydrogen production is projected to increase from the current 0.7 Mt/yr to 125 Mt/yr by 2030 and 523 Mt/yr by 2050.

An analysis of the energy transition as a socio-technical transformation acknowledges that the social, political, and economic structures that influence energy production and consumption are also changing, in addition to the technological advances utilized to produce and consume energy. As a result, the transition will be unique to each nation. In Brazil, electricity is mainly produced from hydropower. Brazil's government highlights that the nation is proud to be a leader in renewable energy. In 2020, electricity energy mix mainly comprised of hydro (60.7%), biomass (9.0%), wind (8.7%), natural gas (9.1%), solar (1.6%), and coal (2.7%). However, in 2021, the share of hydro was reduced to 53.4% due to water scarcity [42]. Moreover, the discovery of oil in the pre-salt layer hindered the country's agenda for renewable energy. The government anticipates the revenue from oil will be used to promote the use of renewable energy [43].

The World Economic Forum released a report on "Fostering Effective Energy Transition" in June 2023 [44]. This report lists the Energy Transition Index (ETI) for 120 countries. The 2023 ETI framework has been revised to include a more comprehensive strategy for striking a balance between the equity, security, and sustainability

Table 3 Energy transition index [44]

Country	ETI rank
Norway	3
USA	12
Brazil	14
China	17
Canada	19
Malaysia	35
Saudi Arabia	57
UAE	63
Mexico	68
Kazakhstan	76
Oman	90
IR Iran	92
Kuwait	102

imperatives of the energy triangle. The report highlighted countries that stood out for laying the groundwork for a robust energy transition. These countries included Brazil, Canada, Nigeria, Saudi Arabia, UAE, and the USA, which are oil-rich countries. The selected oil-rich countries' ETI is listed in Table 3. The best rank was for Norway and the least was for Kuwait. Kuwait lacks the proper energy transition plan and supporting legislation and policies. Kuwait is still highly dependent on oil and gas with minimal use of solar energy.

A strong framework for decision-making has been offered by Alghassab [45] in relation to Saudi Arabia's energy transformation. His analysis covered a wide range of alternative energy sources, including biomass, geothermal, wind, solar, and ocean energy. His investigation revealed that solar energy was the most beneficial option out of all the options taken into consideration for the country's sustainable energy transition. When compared to other renewable energy sources evaluated, solar energy shows significant promise and advantages in terms of economic feasibility, technological feasibility, and environmental sustainability. The findings provide significant direction for stakeholders in Saudi Arabia, enabling them to make educated decisions, expedite their energy transition endeavors, and accomplish sustainable development goals.

Belaïd and Al-Sarihi examined the parameters that influence in molding the modern energy transition in Saudi Arabia using the multi-level perspective (MLP) approach [46]. The multi-level perspective (MLP) approach views transitions as interactions between three distinct but interconnected levels: the landscape, which provides the broader societal context; the regime, which represents the existing system; and the niche, where innovations and new practices emerge. Understanding how these levels interact and influence each other is crucial for analyzing and facilitating transitions in socio-technical systems.

They have opined that the interaction between the internal and external forces shapes Saudi Arabia's energy transition. The external factors included climate treaties enforcement and the general concerns about the possible negative consequences of a rapid global energy transition. On the other hand, the internal factors that they referred to include the escalating national energy consumption and GHG emissions.

They state that Saudi Arabia's modern energy transition one evolved from a cautious approach to a progressive one. They concluded that in the Saudi context, the success of the clean energy transition is conditional on direct government support while the niche and landscape factors had minimal pressure on sociotechnical regimes. Yet, they added that it is important to align landscape, regime, and niche developments to realize an energy transition holistically.

The World Energy Transitions Outlook 2023 calls for bold and transformative actions in the energy transition as follows:

- A significant scale-up of renewables is emphasized, alongside investments in supporting infrastructure.
- Comprehensive policies are crucial for deployment and ensuring broad socio-economic benefits.
- Embedding net-zero commitments in legislation and well-resourced implementation plans is deemed essential.
- Structural changes within three decades are necessary to align with the Paris Agreement goals.
- Long-term considerations include electrification, modernizing infrastructure, and addressing challenges.
- Advocacy for a just and inclusive energy transition to promote human well-being, equity, and sustainability.
- Supporting developing countries in accelerating the transition is vital for global energy security.
- A diverse energy market is encouraged to reduce supply chain risks and ensure local value creation.
- Access to technology, training, capacity building, and affordable finance is essential for unlocking countries' potential in the global energy transition.
- Human welfare and security must remain central to the energy transition, necessitating systemic changes beyond the energy sector to address pervasive issues and inequalities.
- A renewables-based energy transition has the potential to alleviate conditions underlying energy transition challenges.
- Representation and integration of community needs and interests into transition planning are crucial for a successful and accepted energy transition.

Technology Innovation

Samargandi looked into the factors that affect the energy intensity of OPEC nations. The results of this study confirmed that trade openness is a major factor in reducing energy intensity and show that, in contrast to renewable energy, innovation has a negligible relationship with energy intensity [47]. Later in 2023, Samargandi and Lin-chu opined that innovation is also insignificantly associated with energy intensity. While many indicators can capture innovations, such as patent development, in their analysis renewable energy and patents were used as a proxy for innovation in OPEC countries. They opined that their observations using patents as the innovation proxy nullified their hypothesis that "innovation contributes to lowering energy intensity (measured by patent)." Conversely, when renewable energy served as the innovation proxy, the hypothesis that "innovation helps decrease energy intensity" was supported. They concluded that OPEC countries should utilize oil rent to develop technological advancement to enhance energy efficiency [48].

According to Sun et al., technology innovation can advance through both domestic and foreign knowledge [49]. Domestic knowledge refers to innovation efforts within the host country, while foreign knowledge originates abroad. Sun et al. elaborate on how foreign knowledge can impact a domestic economy, highlighting knowledge transfer and unintentional knowledge spillover as potential mechanisms. They emphasized that technological innovation, whether originating domestically or abroad, holds the potential to enhance global energy efficiency. Acknowledging the greater impact of foreign innovation compared to domestic innovation, it is recommended that national policymakers actively foster domestic innovative capabilities and technologies. Likewise, the development of human capital is crucial for effectively leveraging foreign knowledge spillover.

Malaysia, a net exporter of natural gas and oil, is not classified as an oil-rich nation despite its substantial oil reserves. However, exploring its energy utilization and technological advancements provides a valuable illustration of a developing country with ample oil reserves. In this context, Sohag et al. examined the complex relationship between technology innovation and energy use. Their findings highlight that technical innovations play a crucial role in reducing energy consumption by enhancing the productivity of manufacturing operations [50].

Waheed et al., assessed the influence of high technology exports on energy consumption, energy intensity, and carbon intensity in GCC countries spanning from 1990 to 2017 [51]. Examples of high-technology industries are aircraft, computers, and pharmaceuticals; medium-high-technology includes motor vehicles, electrical equipment, and most chemicals. Their findings revealed that high-technology exports contributed to a reduction in overall energy demand and energy intensity. They recommended the implementation of novel and innovative energy conservation policies to encourage the adoption of energy-efficient equipment and technologies, fostering cleaner and greener growth. The World Bank illustrates that the UAE outperforms other selected GCC countries in high-technology exports, as depicted in Fig. 8 [52].

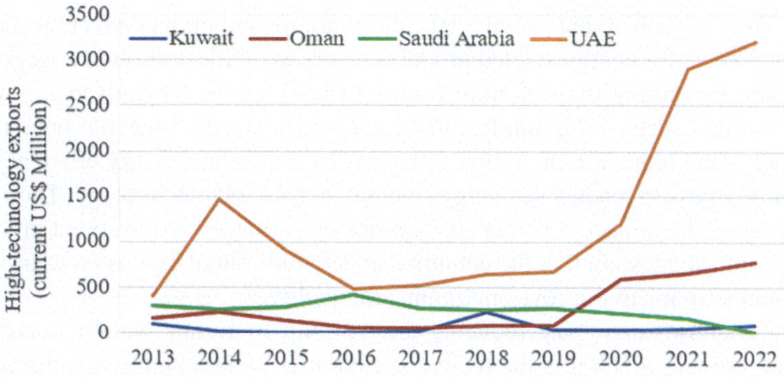

Fig. 8 High technology exports of selected GCC countries (current USA $) [52]

Waheed et al., suggested that policymakers should focus on integrating energy considerations within the design stage of technology products and promoting environment-friendly investments while deciding the export baskets [51]. They also recommended advocating for financial assistance and easing trade tariffs to encourage the advancement of high-tech products like zero-emissions technologies (ZETs), hybrid and electric vehicles, and large-scale clean-energy production. Additionally, diversifying the array of funding instruments for research and development activities is suggested. Furthermore, governments can levy environmental taxes on high-emission industries to compel their adoption of advanced and eco-friendly technologies, contributing to energy conservation and mitigating environmental pollution. From another perspective, Khan et al., evaluated the impact of energy consumption in the global panel from 1980 to 2019 for 179 countries on four innovation proxies [53]. The innovation proxies that were considered included the following:

1. Patent application residents (number per thousand population).
2. Patent application nonresidents (number per thousand population).
3. Research and development expenditure (percentage of real gross domestic product).
4. High-technology exports (percentage of real gross domestic product).

They employed two static models, the System Generalized Method of Moments (GMM), and panel quantile regression in their analysis. The outcomes from both methods were consistent regarding the relationship between energy consumption and innovation proxies. Energy consumption exhibited a negative correlation with all proxies except research and development. Consequently, energy consumption has not yet been shown to contribute to enhancing the innovation level, although it has positively influenced research and development innovation. Similar results were found by H. Khan and I. Khan as they evaluated the effect of technological innovations, urbanization, and economic growth on environmental quality [53]. Their conclusion was that technical advancements and foreign direct investment increase carbon dioxide emissions because they increase energy consumption, yet technical advancements increase research and development which improves environmental quality. They explained that patent applications are filed to protect novel inventions. Securing patents for energy-related products serves as an illustration of engagement in activities promoting innovation in energy technology, thereby advancing developments in this sector. This implies that increased innovation has the potential to contribute to the reduction of carbon emissions by enhancing energy efficiency and facilitating access to renewable energy sources such as solar technology. Their findings regarding the impact of patent applications on carbon emissions point in a positive direction, suggesting that the countries under study should focus on reinforcing their patent systems to improve environmental quality.

Many countries anticipate reducing their carbon footprints without sacrificing economic growth. Since it is the main contributor to carbon emissions, the energy sector has drawn attention from numerous nations worldwide. Moreover, technological innovation is one of many approaches that can reduce energy consumption consequently lowering carbon emissions. In that regard, Erdoğan et al. investigated how innovation affected sectoral carbon emissions for fourteen Group of 20 (G20) nations. Leading industrialized and developing nations participate in the G20 forum. Through the forum, they come together to deliberate and plan strategies regarding matters on global economic policy. The chosen countries for their research included nations for which data was obtainable. A portion of these countries were oil-rich nations, such as Brazil, Canada, Mexico, and the United States. Their analysis spanned over the period from 1991-2017. Due to the differing technological advancements in machinery and equipment used during production, energy consumption varies accordingly. Their findings highlighted the substantial influence of technological innovation within the industrial sector on energy usage as the integration of newer, more energy-efficient technologies significantly impact energy consumption patterns in this domain [54].

In summary, the above-mentioned research studies explored the intricate relationship between technological innovation, energy consumption, and environmental impact. These studies delved into diverse aspects, including the use of proxies like patents and renewable energy to measure innovation's impact on energy intensity. Findings suggested that innovation, particularly through renewable sources, plays a crucial role in reducing energy intensity. The significance of both domestic and foreign knowledge in technological advancement is emphasized, with recommendations for policymakers to foster domestic capabilities. The role of technical innovations in diminishing energy consumption is underscored, offering valuable insights into the complexities of this relationship. Additionally, analyses of high-tech exports' influence on energy consumption and the examination of energy's impact on innovation and sectoral carbon emissions contribute to a comprehensive understanding of the intricate dynamics between energy, innovation, and environmental outcomes.

Key Factors Influencing Energy Intensity

To evaluate the impact of important factors on energy intensity. Some factors were selected based on the above discussions. Accordingly, the following spider (radar) graph (Fig. 9) was formed.

The energy intensity and renewable energy (RE) scale are shown on the energy intensity axis (range 0–12), while all other factors are shown on the rank axis (0–120). Several indicators are included that are believed to affect the energy intensity including the United Nations Sustainable Development Goals (UN SDGs), the United Nations Human Development Index (HDI), and the Energy Transition Index (ETI). Renewable energy (RE) (i.e. solar, wind, biomass ... etc.) is also included as it is a factor affecting countries' energy efficiencies and hence intensities.

Energy intensities for the selected countries as discussed earlier ranged between 3.21 MJ/$2017 PPP GDP and 8.75 MJ/$2017 PPP GDP. The lowest energy intensity was for Mexico. Mexico does not have high ranks in the UN SDG index rank (a lower rank number is better). Yet it has the lowest energy intensity. Mexico's largest industries include tobacco, aerospace, petroleum, and mining [55]. Mexico's natural gas as a source of electricity generation increased from 7 to 62.3% in 1990 and

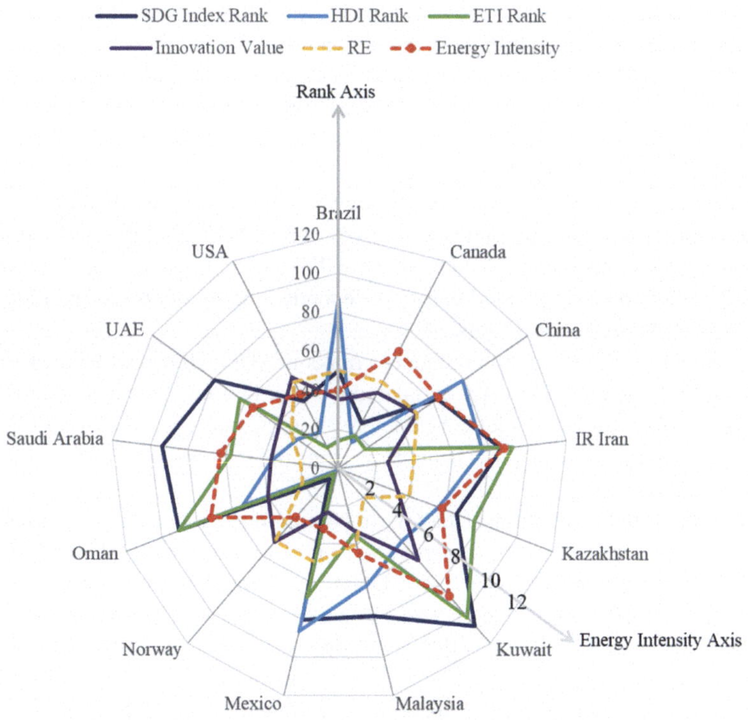

Fig. 9 Radar chart on impact factors of energy intensity

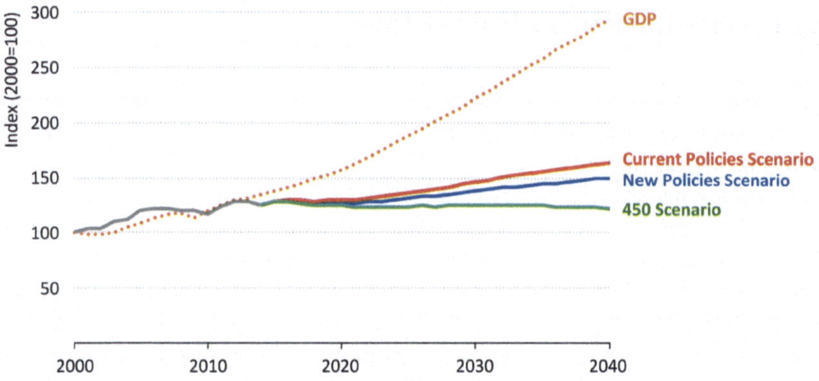

Fig. 10. Primary energy demand and GDP in Mexico by scenario, 2000-2040 (indexed to 2000 level) [57]

2020 respectively [56]. These efforts targeted the reduction of CO_2 emissions. Since 2013, Mexico has undergone energy sector reforms, opening the market to private investment. This led to increased competition, innovation, and more efficient energy use as shown in Fig. 10 [57, 58].

Various scenarios were presented in Fig. 10, including the current policies highlighting the impact of decoupling energy demand from GDP. The realization of this decoupling is attributed to a range of initiatives extending beyond the privatization of the energy sector. These efforts encompass improvements in energy efficiency, particularly within industrial activities, given that the industrial sector constitutes half of the country's electricity demand.

Norway as shown in Fig. 10, has the best portfolio in terms of these selected indicators and ranks. Norway's energy intensity of 3.36 MJ/$2017 PPP GDP. It scores among the top nations in the UN SDG index, HDI, and ETI ranks, with respective of 7, 2, and 3. Norway as explained before is dependent on hydropower and only 20% of its energy mix from fossil fuels.

Canada, while achieving commendable ranks in SDG, HDI, and ETI, exhibits a somewhat higher energy intensity compared to other nations. This can be attributed to its dependence on fossil fuels in its energy mix. Canada has the opportunity to enhance its energy intensity by benefiting from other countries such as Mexico, Brazil, and Malaysia.

In summary, the evaluation of factors impacting energy intensity culminated in a comprehensive spider graph, incorporating indicators such as the United Nations Sustainable Development Goals (UN SDGs), the United Nations Human Development Index (HDI), the Energy Transition Index (ETI), and renewable energy (RE). Notably, various countries displayed diverse energy intensity profiles, with Mexico achieving a particularly low energy intensity despite its relative standing in the UN SDG index. Mexico's success is attributed to post-2013 energy sector reforms that encouraged private investment, fostered competition, and spurred innovation. Conversely, Norway emerged as a leader in energy efficiency, while Canada faces challenges in energy intensity due to a notable reliance on fossil fuels. The study underscores the potential for countries to draw valuable insights from one another, with Canada poised to benefit from the experiences of Mexico, Brazil, and Malaysia in enhancing energy intensity and contributing to global sustainability objectives. These trends reflect the unique energy dynamics of oil-rich countries, shaped by their abundant natural resources and their position in the global energy market.

Discussions and Summary

The complex dynamics that define oil-producing nations and their significant influence on the world energy scene were investigated. One characteristic that sets it apart is its substantial reliance on fossil fuels, especially natural gas and oil, which influence both domestic energy consumption and export-oriented energy output. These countries' economic success is clearly linked to the price of oil globally and the amount of oil they export, which leads to a variety of patterns in energy use.

Interestingly, certain oil-producing nations offer energy subsidies, which encourage artificially low pricing and raise consumption rates as a result. These subsidies in some nations are a way of sharing national wealth. Moreover, the restricted diversity of the energy mix to the increased energy intensity. These countries' susceptibility to supply disruptions and price variations stems from this fragility, which calls for careful assessment in the context of the global energy situation.

Turning to the statistics of oil-rich countries, I appreciate the definition based on substantial crude oil reserves, making them key players in the global oil market. The wealth generated from oil resources, as highlighted, significantly influences economic stability, geopolitical dynamics, and international relations. The inclusion of key players like Saudi Arabia, Russia, the United States, and Canada provides a well-rounded understanding of the geopolitical landscape.

Examining the impact of oil reserves on global oil production, particularly the USA leading in 2022, aligns with my understanding of the complex and dynamic nature of global oil markets. The discussion on the repeal of the crude oil export ban in 2015 and the subsequent historic peak in USA crude exports in 2023 underscores the real-time influence of policy changes on production capacities.

The in-depth examination of the intricate interplay among energy statistics, consumption patterns, and the geopolitical and economic importance of oil-rich nations has established a robust groundwork for subsequent analysis. This is particularly crucial for comprehending technological advancements and the wider implications they carry for global energy transitions.

Oil-rich countries, characterized by their heavy reliance on fossil fuels, particularly oil and natural gas, play a pivotal role in shaping the global energy landscape. Their influence which extends across various dimensions was explored. Several factors were elaborated on including energy supply and global prices, the economic and fiscal impact of foreign exchange reserves, the geo-political influence of energy transitions, infrastructure and investments, global trade and diplomacy, and energy policy shifts. It was explained that the influence of oil-rich countries on the global market is profound, extending beyond energy production and supply to encompass economic, political, and financial dimensions with far-reaching effects on global stability.

The heavy reliance of Oil-rich countries on fossil fuels, particularly oil and natural gas, wields substantial influence across various dimensions of the global energy landscape. As major producers and exporters of petroleum products, disruptions in their oil production, as seen in the 2019 attacks on Saudi Aramco, send immediate ripples through global markets, affecting oil prices and critical sectors such as energy, manufacturing, and transportation. The economies of these nations are intricately tied to oil exports, contributing significantly to GDP and government budgets. Fluctuations in oil prices can lead to economic volatility, impacting global financial markets and necessitating fiscal adjustments. Accumulating significant foreign exchange reserves, often managed through sovereign wealth funds like Norway's Government Pension Fund Global, further shapes international capital flows. Additionally, their geopolitical stability is crucial for global energy security, while investments in renewable energy and sustainable technologies contribute to broader global energy transitions. Globally, oil-rich nations strategically invest in energy infrastructure, influencing logistics, and their energy trade agreements impact international relations and diplomacy. Some are gradually shifting energy policies to reduce dependence on fossil fuels, exemplified by Norway's early adoption of incentives for sustainable transportation. In essence, the influence of oil-rich countries transcends mere energy production, encompassing economic, political, and financial dimensions that resonate globally.

The energy trilemma as a concept that encapsulates the challenges of achieving a secure, equitable, and sustainable energy system was also discussed. As a holistic approach to addressing the energy trilemma, the energy transitions concept was discussed. Energy transitions, envisioned as a holistic strategy, entail a fundamental shift from conventional fossil fuels to sustainable and renewable sources, guided by the World Energy Transitions Outlook 2023. This roadmap aligns with the Paris Agreement, emphasizing increased renewable electricity generation, improved energy intensity, and enhanced clean hydrogen production. Notably, each nation's transition is unique, as seen in Brazil's socio-technical energy evolution and Saudi Arabia's complex interplay of internal and external factors. Thus energy transitions require systematic changes, extending beyond modifications to energy technologies. It involves transformative adjustments within the overarching frameworks of politics, society, the environment, and the economy that revolve around the processes of energy production and consumption.

The World Economic Forum's Energy Transition Index recognizes countries making significant strides in balancing equity, security, and sustainability imperatives, with oil-rich nations like Saudi Arabia and the UAE featured prominently. The World Energy Transitions Outlook 2023 advocates for bold and transformative actions, emphasizing the scale-up of renewables, comprehensive policies, legislative commitments, and structural changes. It underscores long-term considerations such as electrification, infrastructure modernization, and the imperative for a just and inclusive transition. Supporting developing countries, prioritizing human welfare, and ensuring community representation emerge as critical aspects for the global energy transition's success and widespread acceptance.

Exploring the intersection of technology innovation, energy consumption, and environmental sustainability reveals intricate relationships. The impact of innovation on energy intensity is nuanced, with findings indicating that while trade openness reduces energy intensity, traditional measures like patents show a negligible connection. However, using renewable energy as a proxy for innovation suggests a potential decrease in energy intensity. Acknowledging the importance of both domestic and foreign knowledge, policymakers are urged to nurture domestic capabilities to drive technological innovation. Various studies emphasize the role of technical innovations in reducing energy consumption, especially in developing nations and G20 countries. Collectively, these insights shed light on the complex dynamics involved in achieving a sustainable energy system through technological advancements.

The assessment of factors influencing energy intensity culminates was summarized in a comprehensive spider graph, incorporating indicators like the United Nations Sustainable Development Goals (UN SDGs), the United Nations Human Development Index (HDI), the Energy Transition Index (ETI), and renewable energy (RE). Notably, diverse energy intensity profiles are observed among countries, with Mexico achieving a remarkably low energy intensity despite its standing in the UN SDG index. This success is attributed to post-2013 energy sector reforms that stimulated private investment, competition, and innovation. Conversely, Norway emerges as an energy efficiency leader, while Canada grapples with higher energy intensity due to a notable reliance on fossil fuels. As a result, there is a good chance that nations will share knowledge, with Canada in an excellent position to gain from Mexico, Brazil, and Malaysia's experiences in raising energy intensity and supporting international sustainability goals.

Conclusions

This chapter offered an insightful exploration into the complexities of energy consumption in oil-producing nations, emphasizing the pivotal role of the energy mix and technology in shaping their patterns of energy use. The examination shed light on the intricate dynamics of these nations' substantial global influence, considering economic, geopolitical, and environmental dimensions. The energy trilemma concept underscored the challenges in achieving a secure, equitable, and sustainable energy system, with the proposed energy transitions approach providing a roadmap for transitioning towards renewable sources. The World Economic Forum's Energy Transition Index acknowledged strides in balancing equity, security, and sustainability, prominently featuring oil-rich nations.

Additionally, the analysis of technology innovation, energy consumption, and environmental sustainability highlighted the nuanced impact of innovation on energy intensity. While trade openness reduced energy intensity, conventional patent measures showed negligible connections. However, the potential of renewable energy to drive down intensity emphasized the need for policymakers to foster domestic innovation capabilities. Recommendations included actively promoting domestic innovative capabilities, developing human capital for leveraging foreign knowledge spillover, and integrating energy considerations in technology design. Technical innovations, particularly in manufacturing, were recognized for their role in reducing energy consumption. Policymakers were urged to focus on environmentally friendly investments and considerations in export decisions.

The assessment of factors influencing energy intensity, presented in a spider graph, unveiled diverse profiles among nations. Mexico's success in achieving low energy intensity post-2013 reforms suggested the potential for shared knowledge. Conversely, Norway excelled in energy efficiency, while Canada faced higher intensity due to fossil fuel reliance. This comprehensive analysis, coupled with recommendations for fostering innovation and sustainability, establishes a solid foundation for further exploration. The emphasis on international collaboration underscores the importance of raising energy intensity and supporting sustainability goals globally.

Leveraging the insights gained from the intricate examination of energy statistics in oil-rich countries, decision-makers can make informed and strategic decisions for navigating the complexities of the global energy landscape.

References

1. OPEC Annual Statistical Bulletin (ASB), 2023. https://asb.opec.org/data/ASB_Data.php. Last accessed October 12, 2023.
2. GAO, "Crude Oil Markets: Effects of the Repeal of the Crude Oil Export Ban," Washington, DC, USA, 2020. [Online]. Available: https://www.gao.gov/products/gao-21-118#:~:text=PubliclyReleased%3ANov20%2C 2020, of U.S. crude oil worldwide.
3. EIA, 2023. U.S. Energy Information Administration (EIA), "Today in Energy U.S. crude oil exports reached a record high in first half of 2023," *Today in Energy*, Washington, DC, USA, Oct. 2023.
4. Krane J. and Finley M., 2019. The US Response to Attacks on Persian Gulf Oil Infrastructure and Strategic Implications for Petro-States. Rice University Baker Institute for Public Policy. Issue brief. www.bakerinstitute.org/issue-briefs. Last accessed October 20, 2023.
5. MacroTrends, 2023. Brent crude oil prices. https://www.macrotrends.net/2480/brent-crude-oil-prices-10-year-daily-chart. Last accessed October 12, 2023.
6. EIA, 2021. U.S. Energy Information Administration (EIA), Crude oil prices peaked in early 2012. https://www.eia.gov/todayinenergy/detail.php?id=7630. Last accessed October 13, 2023.
7. Altalhi, Hassen Hussain, 2023. Transitioning Away from an Oil-based Economy through Saudi Arabia's Vision 2030. Scientific Journal of King Faisal University, Humanities & Management Sciences. 2023, Vol. 24 Issue 2, p41-48. 8p.
8. Dymitrowska, Yanina, 2023. Savings natural resource funds: effectiveness of the Norwegian Government Pension Fund Global. Economics and Law, Vol. 22, no. 3, pp. 471-495.
9. Olver-Ellis, Sophie, 2020. Building the new Kuwait: Vision 2035 and the challenge of diversification. LSE Middle East Centre Paper Series (30). LSE Middle East Centre, London, UK.

10. Jraissati Jorge, Jakee Keith, 2022. Venezuela's collapse: Exogenous shock or institutional design? Economic Affairs, Volume 42, Issue 2. Pages 344-360. https://doi.org/10.1111/ecaf.12537.
11. S. Resaland, L. A. Riis, and S. K. Mohn, "Government Pension Fund Global's investments in petroleum equities," NORWEGIAN SCHOOL OF ECONOMICS, 2015.
12. R. Abdelal, "Sovereign Wealth in Abu Dhabi," *Geopolitics*, vol. 14, no. 2, 2009, https://doi.org/10.1080/14650040902827781.
13. M. Nerlinger and S. Utz, "The impact of the Russia-Ukraine conflict on energy firms : A capital market perspective," *Financ. Res. Lett.*, vol. 50, no. May, p. 103243, 2022, https://doi.org/10.1016/j.frl.2022.103243.
14. Hook L. and Hume N., 2022. Will the Ukraine war derail the green energy transition? Financial Times, last accessed on October 1, 2023. https://www.ft.com/content/93eb06ec-ba6c-4ad2-8fae-5b66235632b2.
15. Hancock A., 2023. Ukraine war and Biden's IRA force EU to accelerate energy transition. Financial Times, last accessed on October 15, 2023. https://www.ft.com/content/eb597d26-fd02-4ca0-94fc-74bd17e0f3b3.
16. Directorate-General for Energy (DGoE-a), "State of the Energy Union Report 2023," Brussels, 2023.
17. Directorate-General for Energy (DGoE-b), "Assessment of progress towards the objectives of the Energy Union and Climate Action highlights the need for EU countries to focus on implementation," 2023. https://energy.ec.europa.eu/news/assessment-progress-towards-objectives-energy-union-and-climate-action-highlights-need-eu-countries-2023-10-26_en (accessed Oct. 23, 2023).
18. IEA, 2023-a. International Energy Agency (IEA), "Denmark's oil and gas phase-out," 2023. https://www.iea.org/policies/17708-denmarks-oil-and-gas-phase-out (accessed Oct. 30, 2023).
19. Kingdom of Saudi Arabia (KSA), "Vision 2030." https://www.vision2030.gov.sa/media/rc0b5oy1/saudi_vision203.pdf (accessed Oct. 13, 2023).
20. M. Al-Saidi, "Energy transition in Saudi Arabia: Giant leap or necessary adjustment for a large carbon economy?," *Energy Reports*, vol. 8, pp. 312–318, 2022, https://doi.org/10.1016/j.egyr.2022.01.015.
21. S. C. Singh and N. Verma, 2023. India wants Saudi Aramco to develop strategic petroleum reserve as ties strengthen. https://www.reuters.com/markets/commodities/india-wants-saudi-aramco-develop-strategic-petroleum-reserve-ties-strengthen-2023-10-11/.
22. QatarEnergy ANNUAL REVIEW, 2022 https://www.qatarenergy.qa/en/MediaCenter/Publications/QatarEnergy%20Annual%20Review%202022%20-%20English.pdf
23. Y. Al-Abdullah, F. Al-Ragom, O. Alsayegh, A.-A. Shareefa, K. Mohammad, and K. J. Sreekanth, "Kuwait Energy Outlook 2023," Kuwait, 2023.
24. Kuwait Foreign Petroleum Exploration Company (KUFPEC), "KUFPEC Portfolio," 2023. https://www.kufpec.com/en/portfolio/naar (accessed Oct. 30, 2023).
25. D. Quint and F. Venditti, "The Influence of OPEC+ on Oil Prices: A Quantitative Assessment," *Energy J.*, vol. 44, no. 01, pp. 173–186, 2023, https://doi.org/10.5547/01956574.44.4.dqui.
26. R. Ulatowski, "OPEC+ as a new governor in global energy governance," *Rev. UNISCI*, vol. 2020, no. 53, pp. 241–263, 2020, https://doi.org/10.31439/UNISCI-94.
27. Norsk elbilforening, "Norwegian EV policy," 2023. https://elbil.no/english/norwegian-ev-policy/ (accessed Nov. 01, 2023).
28. A. Yang, C. Liu, D. Yang, and C. Lu, "Electric vehicle adoption in a mature market: A case study of Norway," *J. Transp. Geogr.*, vol. 106, 2023, https://doi.org/10.1016/j.jtrangeo.2022.103489.
29. EAFO, 2023. European Alternative Fuels Observatory (EAFO), "Norway Incentives and Legislation," 2023. https://alternative-fuels-observatory.ec.europa.eu/transport-mode/road/norway/incentives-legislations#:~:text=Norway bases circulation taxes on, vehicles and plug-in hybrids. (accessed Nov. 12, 2023).
30. IEA, 2023-b. International Energy Agency (IEA), Is energy intensity an energy efficiency indicator? https://www.iea.org/data-and-statistics/data-tools/energy-end-uses-and-efficiency-indicators-data-explorer. Last accessed October 31, 2023.

31. Iran Chamber Newsroom, "Iran has world's biggest per capita energy consumption," 2023. https://en.otaghiranonline.ir/news/45199 (accessed Dec. 15, 2023).
32. World Bank-a, "Angola," 2023. https://data.worldbank.org/country/AO (accessed Dec. 15, 2023).
33. MEWRE, "Electrical Energy Statistical Year Book," Kuwait, 2022. [Online]. Available: https://www.mew.gov.kw/media/0fpnlm2h/electrical-annual-report-2022-for-qr.pdf.
34. OECD, "Organization for Economic Co-operation and Development," 2023. https://www.oecd.org/about/members-and-partners/ (accessed Dec. 05, 2023).
35. World Bank-b, "Energy Intensity," 2023. https://data.worldbank.org/indicator/EG.EGY.PRIM.PP.KD (accessed Nov. 03, 2023).
36. Energy Institute Statistical Review of World Energy (2023), OurWorldInData.org/energy-mix, CC BY. https://ourworldindata.org/energy-mix. Last accessed October 20, 2023.
37. Y. Keho, "What drives energy consumption in developing countries ? The experience of selected African countries," *Energy Policy*, vol. 91, no. December 2015, pp. 233–246, 2016, https://doi.org/10.1016/j.enpol.2016.01.010.
38. A. V Komarova, I. V Filimonova, and A. A. Kartashevich, "Energy consumption of the countries in the context of economic development and energy transition," *Energy Reports*, vol. 8, no. May, pp. 683–690, 2022, https://doi.org/10.1016/j.egyr.2022.07.072.
39. M.-J. Nadeau, Y. D. Kim, N. Baikeli, K.-D. Barbknecht, and L. Birnbaum, "2015 Energy Trilemma Index Benchmarking the sustainability of national energy systems," London, 2015. [Online]. Available: https://www.worldenergy.org/assets/downloads/20151030-Index-report-PDF.pdf.
40. ICPAC. IGAD Climate Prediction and Applications Centre (ICPAC). "Energy and Climate : The Dilemma, Trilemma, and Quadrilemma," 2020. https://icpac.medium.com/energy-and-climate-the-dilemma-trilemma-and-quadrilemma-839a8d657369 (accessed Dec. 10, 2023).
41. IRENA, *World Energy Transitions Outlook: 1.5°C Pathway*. 2022.
42. D. Werner, L. Luz, and B. Lazaro, "The policy dimension of energy transition : The Brazilian case in promoting renewable energies (2000–2022)," *Energy Policy*, vol. 175, no. February, p. 113480, 2023, https://doi.org/10.1016/j.enpol.2023.113480.
43. L. L. B. Lazaro, R. S. Soares, C. Bermann, F. M. A. Collaço, L. L. Giatti, and S. Abram, "Energy Research & Social Science Energy transition in Brazil : Is there a role for multilevel governance in a centralized energy regime ?," *Energy Res. Soc. Sci.*, vol. 85, no. November 2021, p. 102404, 2022, https://doi.org/10.1016/j.erss.2021.102404.
44. WEF, "Fostering Effective Energy Transition," 2023. [Online]. Available: https://www.weforum.org/publications/fostering-effective-energy-transition-2023/.
45. M. Alghassab, "A Computational Case Study on Sustainable Energy Transition in the Kingdom of Saudi Arabia," *Energies*, vol. 16, 2023, https://doi.org/10.3390/en16135133.
46. F. Belaïd and A. Al-Sarihi, "Research in International Business and Finance Saudi Arabia energy transition in a post-paris agreement era : An analysis with a multi-level perspective approach," *Res. Int. Bus. Financ.*, vol. 67, p. 102086, 2024, https://doi.org/10.1016/j.ribaf.2023.102086.
47. N. Samargandi, "Energy intensity and its determinants in OPEC countries," *Energy*, vol. 186, p. 115803, 2019, https://doi.org/10.1016/j.energy.2019.07.133.
48. N. Samargandi and L. Lin-chu, "Energy intensity and its determinants in OPEC countries," *Energy*, vol. 186, p. 115803, 2019, https://doi.org/10.1016/j.energy.2019.07.133.
49. H. Sun, B. K. Edziah, A. K. Kporsu, S. A. Sarkodie, and F. Taghizadeh-Hesary, "Energy efficiency: The role of technological innovation and knowledge spillover," *Technol. Forecast. Soc. Change*, vol. 167, p. 120659, 2021, https://doi.org/10.1016/j.techfore.2021.120659.
50. Sohag, K., Begum, R. A., Abdullah, S. M. S., & Jaafar, M. (2015). Dynamics of energy use, technological innovation, economic growth and trade openness in Malaysia. *Energy, 90*, 1497-1507.
51. Waheed, R., Sarwar, S., & Mighri, Z. (2021). Role of high technology exports for energy efficiency: Empirical evidence in the context of Gulf Cooperation Council countries. *Energy & Environment, 32*(5), 803-819.

52. World Bank-c, "High-technology exports," 2023. https://data.worldbank.org/indicator/TX.VAL.TECH.CD (accessed Dec. 05, 2023).
53. H. Khan and I. Khan, "The effect of technological innovations, urbanization and economic growth on environmental quality : does governance matter ?," *Front. Environ. Sci.*, no. October, pp. 1–19, 2023, https://doi.org/10.3389/fenvs.2023.1239288.
54. Erdoğan, S., Yıldırım, S., Yıldırım, D. Ç., & Gedikli, A. (2020). The effects of innovation on sectoral carbon emissions: Evidence from G20 countries. *Journal of environmental management*, *267*, 110637.
55. J. Chepkemoi, "What Are The Biggest Industries In Mexico?," *Economics*, 2018. https://www.worldatlas.com/articles/what-are-the-biggest-industries-in-mexico.html#:~:text=Tobacco%2Caerospace%2Cpetroleum%2C and, some of Mexico's biggest industries. (accessed Dec. 23, 2023).
56. M. S. Vera, L. G. Manrique, I. R. Peña, and A. D. L. V. Navarro, "Drivers of electricity GHG emissions and the role of natural gas in Mexican energy transition," *Energy Policy*, vol. 173, 2023, https://doi.org/10.1016/j.enpol.2022.113316.
57. World Energy Outlook (WEO), "Mexico Energy Outlook," PA, 2016. [Online]. Available: https://iea.blob.core.windows.net/assets/e911d7c8-955a-4495-ace0-da85ac4484f9/MexicoEnergyOutlook.pdf.
58. H. Haberl, D. Wiedenhofer, and D. Vir, "A systematic review of the evidence on decoupling of GDP, resource use and GHG emissions, part II : synthesizing the insights," *Environ. Res. Lett.*, 2020, https://doi.org/10.1088/1748-9326/ab842a.

Drivers of Human Behavior to Meet UN Sustainable Development Goals (SDGs) in Relation With Energy

Neethu Roy and Alice Thomas

Abstract The pursuit of Sustainable Development Goals (SDGs) in the context of energy requires a comprehensive understanding and strategic influence on the key factors involved. The COP26 negotiations at the Glasgow Climate Conference has emphasized the urgent necessity to restrict global warming, cease financing for fossil fuels, and achieve zero emissions by 2050. Human behavior plays a pivotal role in driving progress toward achieving the United Nations (UN) Sustainable Development Goals (SDGs), particularly concerning energy-related initiatives. The pursuit of sustainable energy practices is intrinsically tied to both individual and collective human behaviors shaped by various factors. This article emphasizes the role of individuals in achieving the desired targets of energy sustainability and the major hindrances. Enabling a sustainable energy transition to mitigate anthropogenic climate change requires profound shifts in human perceptions, preferences, and behavior. Successfully achieving these changes necessitates a significant contribution from social scientists, who play a crucial role in comprehending how to motivate and empower individuals to actively participate in a sustainable energy transition.

Keywords Sustainable development goals · Energy transition · Human behaviour · Behavioral choice

Introduction

The concept of sustainable development originates from the ecosystem's capacity to sustain human life. Its objective is to guide human behavior and actions towards a future secured by the renewability of vital natural resources crucial for the well-being of the ecosystem shared between humans and nature. As economies become more

N. Roy (✉) · A. Thomas
Department of Civil Engineering, Mar Baselios College of Engineering and Technology, Thiruvananthapuram, Kerala, India
e-mail: neethu.roy@mbcet.ac.in

A. Thomas
e-mail: alice.thomas@mbcet.ac.in

K. J. Sreekanth (ed.), *Energy Efficiency Improvements with Emission Abatement for Energy Sustainability*, https://doi.org/10.1007/978-981-97-9480-5_6

complex, the necessity for knowledge and technology becomes paramount for their transition to environmentally sustainable practices.

Energy systems play a crucial and multifaceted role in human societies, influencing and being influenced by various aspects of daily life, cultural dynamics, and global structures [1]. Consequently, energy is fundamental to the functioning of vital infrastructures such as food, water, communication, transport, manufacturing, security, housing, and much more. It is at the core of the relationships between people and technology that have developed, evolved, and intertwined over centuries [1].

The growth-energy relation was predominantly dependent on energy sourced from slow and inefficient processes such as "plant photosynthesis" before the Industrial Revolution, which drastically changed following the introduction of fossil fuels. While the widespread adoption of fossil fuels brought about a revolutionary transformation in economies, providing unprecedented energy resources for industrial growth and technological advancement, it came at a significant environmental cost. The combustion of fossil fuels, including coal, oil, and natural gas, has released significant quantities of greenhouse gases into the atmosphere, exacerbating climate change. This shift has led to rising global temperatures, more frequent and severe weather events, disruptions to ecosystems, and numerous other climate-related impacts.

The United Nations Sustainable Development Goals (UN SDGs) place significant emphasis on energy-related targets as a critical component of global sustainable development. These goals specifically aim to ensure access to affordable, reliable, sustainable, and clean energy, which is crucial for the development of agriculture, business, communications, education, healthcare, and transportation (SDG 7). The emphasis on energy within the UN SDGs reflects the recognition that sustainable energy practices are fundamental to addressing various interconnected challenges, including poverty, health, economic growth and climate change.

Impacts of climate change being more apparent in this century, innovation is expected to be crucial in driving national and subnational decarbonization efforts [2]. Recent advancements provide both the private and public sectors with a range of technological strategies, from low or zero-carbon technologies to reduce greenhouse gas (GHG) emissions, to innovations in carbon capture and storage aimed at combating the effects of global warming [1]. However, the field of innovation studies, which includes diverse approaches such as technology and innovation management and innovation policy, acknowledges that addressing global challenges through innovation requires not only technological advancements but also organizational, social, and economic changes.

Matos et al. [2] has reviewed many papers on technological innovations and the associated climate change [2]. The paper has highlighted that research on innovation has not adequately delved into three critical subjects: (i) integrating behavioral change with technological innovation; (ii) understanding the socio-technical drivers that expedite low-carbon transitions, and (iii) exploring the role of digital technologies as emerging avenues for addressing managerial challenges in mitigating climate change. Also, research has indicated that the transition of new technology to everyday use involves a co-evolving process involving technology, regulations, infrastructure, and consumer behavior. Policymakers are also seeking cost-effective and sustainable approaches to address the increasing global energy needs. While traditional policies have primarily emphasized supply-side interventions to promote energy efficiency through green innovation and new technologies, numerous studies suggest that relying solely on supply-side policies may not be sufficient to bring about the changes necessary for a sustainable future [3].

This chapter explores the pivotal role of human behavior in the adoption of new technologies for sustainable energy usage. It delves into various research articles that discuss the key drivers of human behavior contributing to the achievement of sustainable development goals related to energy.

Role of Individuals in Attaining SDGs

The significance of sustainable practices extends beyond production, as highlighted by Fujii and Managi [4], and encompasses consumption patterns as well. While global governments have been advocating for the transition to renewable energy to mitigate carbon and greenhouse gas emissions [5], there is an increasing emphasis on adopting environmentally friendly consumption lifestyles in recent years. This dual focus on both sustainable energy production and green consumption reflects a comprehensive approach to address environmental challenges (e.g., [6, 7]). Households contribute to emissions in two main ways: directly, through activities such as cooking, heating, and transportation, and indirectly, through the consumption of goods whose production processes generate emissions. Research indicates that when both direct and indirect emissions are combined, the share of total emissions attributable to households becomes significantly substantial [6].

Individuals, as proactive agents of change, play a vital role in the broader societal effort to achieve SDGs related to energy by fostering a culture of sustainability, promoting responsible consumption, and advocating for transformative practices at various levels of society. Some of the key aspects of the role individuals are as follows:

(a) Energy conservation: Individuals can contribute by adopting energy-efficient practices in their homes, workplaces, and daily activities. Small steps such as utilizing energy-efficient appliances and fine-tuning heating and cooling systems can add up to significant energy conservation.
(b) Awareness and education: Being informed about the environmental impact of energy choices is crucial. Individuals can actively seek information, educate themselves, and spread awareness within their communities about the importance of sustainable energy practices, renewable resources, and the overall impact on the environment.
(c) Advocacy for renewable energy: Individuals can advocate for and support the adoption of renewable energy sources both at the community and policy levels. This may involve promoting clean energy technologies and encouraging businesses and governments to invest in sustainable energy solutions.

(d) Responsible consumption: Making sustainable choices in terms of consumption patterns including opting for products and services that prioritize energy efficiency, are produced using eco-friendly methods, and contribute to a circular economy.
(e) Community engagement: Engaging with local communities and participating in grassroots initiatives focused on sustainable energy can amplify individual efforts. Collaborative projects and shared energy solutions may empower individuals to contribute collectively to SDGs.
(f) Policy Advocacy: Individuals can actively engage in advocacy efforts to influence policies related to energy and the environment. This involves participating in discussions, supporting initiatives that promote sustainable energy practices, and holding policymakers accountable for adopting eco-friendly policies.

Achieving a sustainable energy transition necessitates comprehensive changes in various aspects of energy behavior. This includes the widespread adoption of sustainable energy resources and energy-efficient technologies, substantial investments in energy efficiency measures within buildings, and significant alterations in overall energy use behavior. It is imperative to consider not only the direct consumption of energy but also the indirect energy use associated with the entire life cycle of products, encompassing their production, transportation, and disposal [8].

Limits of Supply-Side Interventions Alone

The scientific community has widely acknowledged the risks associated with escalating energy consumption. However, there has been a prevailing tendency to perceive issues related to energy consumption, greenhouse gas emissions and climate change primarily as engineering challenges that can be solved solely through technological advancements [3]. While technological innovations can be pivotal in achieving sustainable energy consumption by reducing energy input and transitioning to clean, renewable sources, it's recognized that relying solely on technological solutions is inadequate. Despite continual technological advancements progress, global energy demands have not exhibited signs of decline. Technological enhancements might lead to using fewer inputs for a given output, but they could trigger complex reactions, encouraging producers to increase energy usage or consumers to demand more energy-intensive goods. Over time, economies of scale might encourage producers to expand production, resulting in higher energy demand. If this increased production coincides with falling prices, it may further drive-up consumer demand and exacerbate energy consumption pressures [3].

Based on a study of firms operating in the European energy sector, it is reported that sustainable performance is positively associated with firms' financial performance and sustainability reporting is an important driver of sustainable performance [9].

Behavioral Aspects in Energy Modeling

Digital models are vital for assessing the future paths of the energy transition and are commonly employed by decision-makers. However, these models tend to prioritize techno-economic aspects, neglecting in-depth representation of social aspects [10]. Given the increasing importance achieving a fair transition, integrating non-technical factors and limitations into energy models becomes crucial, enhancing the modeling process as a whole. Krumm et al. [11] has studied the potential of various models, namely, energy system models (ESM), integrated assessment models (IAM), agent-based models (ABM), and computable general equilibrium (CGE) models. Details of all these modeling techniques are available in Krumm et al. [11]. Researchers explore various aspects of socio-demographics, societal patterns, behavioral changes, social attitudes, and social innovations in the context of energy. These interconnected levels encompass lifestyle, social movements, technology acceptance, energy citizenship, and innovative experiments, potentially leading to new configurations within energy systems [10].

Krumm et al. [11] focussed on investigating behavior and lifestyle patterns of individuals (primarily consumers), the diverse characteristics among individuals, public attitudes including acceptance and resistance, involvement and ownership of the public, and the dynamics of transformation. Other important aspects of behaviour are the awareness of environmental problems and the contribution of each one towards the same, subjective attitude, altruistic and reciprocal behaviour, health and happiness, and enthusiasm of younger generation [6, 12, 13].

(a) The aspect of behavior and lifestyle includes both material and non-material elements, such as the needs, values, norms, and preferences of consumers. Changes in the behavior of individuals lead to shifts in demand, thereby influencing the requisite development and allocation of renewable energy resources.
(b) The diversity among consumers and producers at various stages of the energy transition, along with the behaviors exhibited by these groups, affects the pace of the transition [14]. This encompasses contextual and environmental factors, the distributional effects of environmental changes and policies, socio-economic circumstances, and the presence of both established entities and innovators.
(c) The rate at which renewable energy is deployed is significantly influenced by public acceptance or opposition toward energy infrastructure. Acceptance, characterized by a positive response to new technology or socio-technical systems, includes three facets: socio-political acceptance (the overall public support for transitioning towards sustainable energy), endorsement from local communities (the approval of specific landscape choices by the local population), and market acceptance (the response of the market to innovative developments). Additionally, this social aspect considers the influence of employment opportunities and local economic growth on individuals' perceptions toward the energy transition [15, 16].

(d) Community approval, particularly when rooted in public engagement and ownership, serves as a vital catalyst for the energy transition. It empowers individuals to shape and actively engage in the local energy transition, fostering a sense of ownership and involvement [17]. Participation can take various forms, such as financial involvement through money transfers, project execution and the decision-making procedures concerning infrastructure initiatives.
(e) Transformation dynamics deals with the transformation at various scales and over time. It encompasses the pace of changes, the influence of historical trajectories, and the characteristics of various system conditions [14].
(f) Awareness—Raising awareness about environmental issues doesn't always result in actual pro-environmental actions. Increased income typically correlates with higher energy usage. Often, the conveniences gained from energy consumption take precedence over concerns about its environmental consequences.
(g) Altruistic and reciprocal behaviour—Various theories, such as those related to preferences and inequality aversion propose that individuals consider the well-being of others in society. The altruistic model suggests that individuals may prioritize the utility of others and be willing to incur some cost to benefit the broader community. In the context of energy consumption, where the negative environmental impacts affect the entire society, adopting reduction behaviors can be viewed as altruistic [6].
(h) Health and happiness—more energy consumption/less energy constraints can generate utility directly by improving the health status of a household [6].
(i) Enthusiasm for innovative technologies and sustainable practices—The enthusiasm of young generation towards climate-resilient development play a vital role in achieving the Sustainable Development Goals (SDGs), as, compared to older generations, young people are more willing to adopt new practices or technologies and are more likely to challenge certain norms that may be hindering sustainable socioeconomic progress [13].

Krumm et al. [11] suggests that the different types of models integrate only certain aspects of the human behaviour and a necessity exists for improved integration of social factors in energy models, and this could be achieved through the integration of diverse model types and the conduct of interdisciplinary research.

Certain studies have focused on making conclusions related to awareness and pro-environmental behaviours [6]. Li [6] explicitly introduced environmental awareness into the altruistic framework to elucidate the environmentally conscious consumption behavior of households. The foundational model posits that each unit of goods consumed entails a detrimental environmental impact. Moreover, the utility is divided into two components: utility derived from health-related consumption and utility associated with other types of consumption.

The basic model by Xiangyu Li is explained in Fig. 1. When accounting for the adverse environmental effects, the household's overall utility (U) exhibits an inverse U-shaped curve. Instead of an insatiable desire to consume without regard for environmental consequences, households will limit their consumption (c) to an optimal

Fig. 1 Utility curve [6]

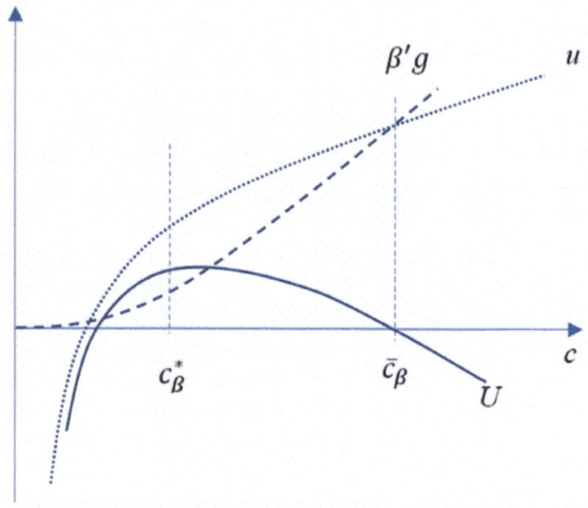

level c^*_β. Once the consumption exceeds maximal level $\underline{c}_\beta$, , additional consumption diminishes the household's utility. The research investigated three common scenarios: (i) where the individual is unaware of the adverse environmental effects and makes decisions solely focused on personal consumption benefits, (ii) where the individual is fully conscious of the environmental impact of their consumption, and (iii) where the individual has partial awareness of their consumption habits and their environmental consequences.

The different hypothesis tested in the study by Li [6] indicate the following:

Awareness can lead to either reduced or increased energy consumption, contingent upon individuals' living circumstances. High-income households might opt for lower energy usage as their environmental awareness grows. It's not just affluent households; even those with lower incomes tend to invest in energy-efficient amenities once they recognize the adverse environmental effects. Higher levels of environmental awareness prompt households to acquire more energy-efficient facilities. A nation wide survey was conducted in China as part of the study (China General Social Survey). Status of Health was used to measure the physical utility generated from energy consumption, while happiness was used to capture the psychological side of the utility, with various control factors such as household income, urban or rural residence, education level, homeownership, car ownership, and marital status were considered. A positive correlation between energy consumption and health/happiness is evident, though the association with happiness appears to exhibit more curvature (Fig. 2). Unhappy and unhealthy households are more likely to have higher energy constraints, although this relationship is nonlinear. (Fig. 3).

Steinberger et al. [18] has drawn categorical differentiations among resource extraction (including primary energy and carbon emissions), economic activity (measured by GDP per capita), human needs satisfiers (such as food supply or household electricity), and well-being outcomes (such as life expectancy) [18]. In

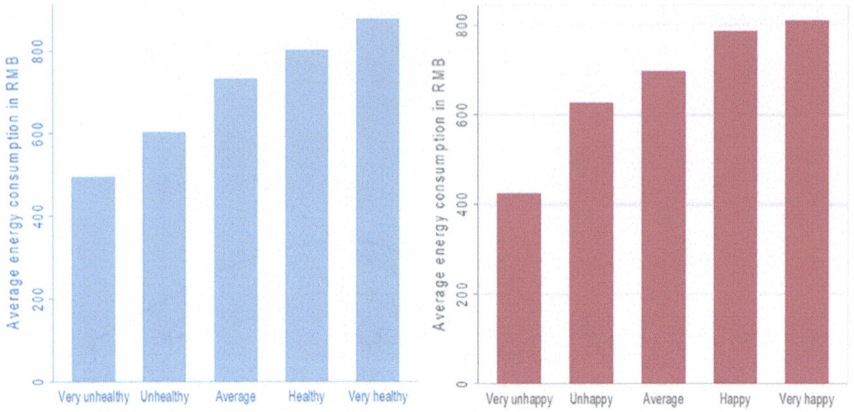

Fig. 2 Electricity consumption and health (left)/happiness (right) [6]

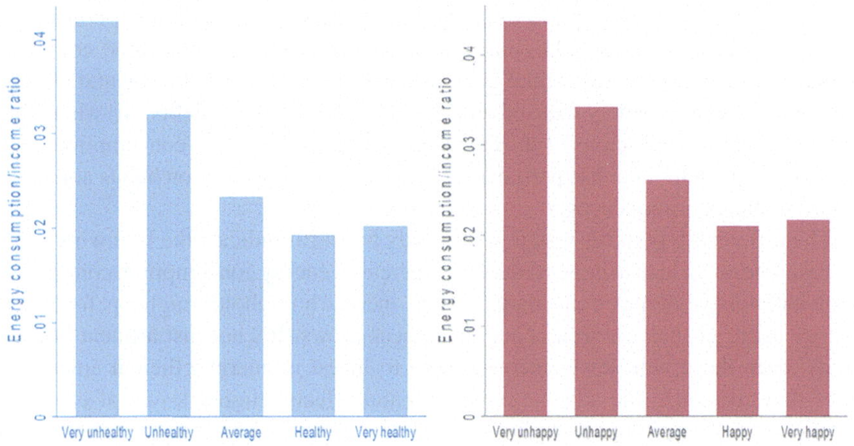

Fig. 3 Electricity consumption/income ratio and health (left)/happiness (right) [6]

Fig. 4, human need satisfiers are separate from well-being outcomes. The former is hypothesized to include final energy as an indicator of the energy services available.

The analysis by Steinberger et al. [18] indicates that the rise in residential electricity consumption statistically explains 60% of the improvements in life expectancy. Additionally, the study highlights that:

PPP Income Growth: There is a strong link between PPP income growth and increases in residential electricity usage.

MER Income Growth: The connection between MER income growth and residential electricity usage is relatively weaker [18] (Fig. 5).

Each arrow indicates the percentage of the change in the destination variable that can be statistically accounted for by the change in the origin variable [18].

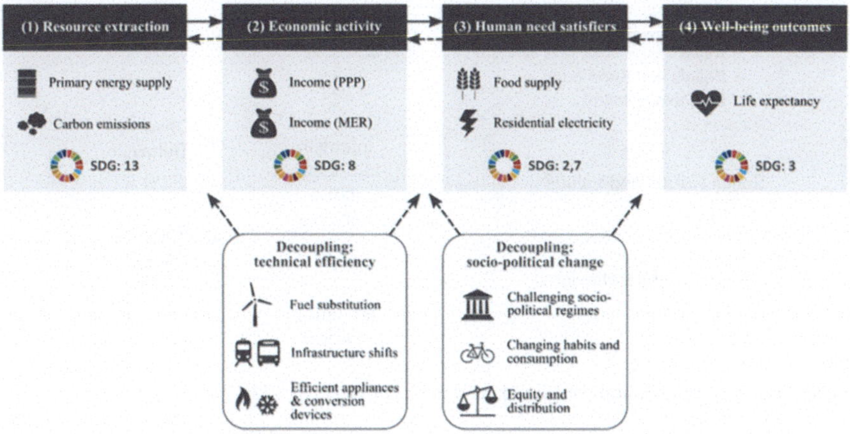

Fig. 4 Analytic framework linking resource extraction and human well-being Income (PPP)—purchasing power parity; Income (MER)—international market exchange rate [18].

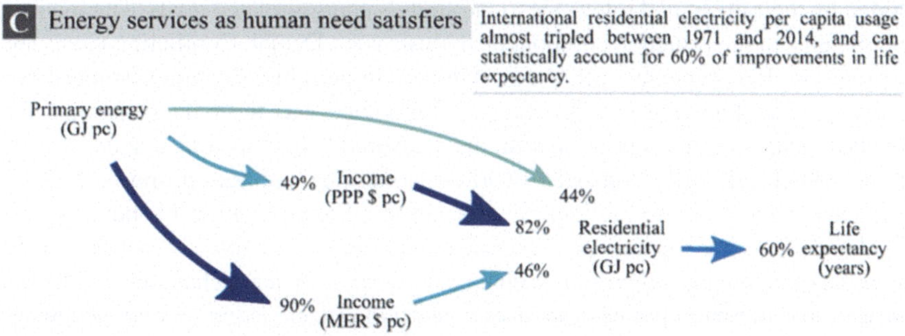

Fig. 5 Testing satisfiers of human needs using functional dynamic decomposition

Theory of Behavioral Choice

It is interesting to note that a significant portion of research in environmental psychology which focused on individuals' environmental awareness, concern, and values, have shown an inconsistent and weak association with pro-environmental behavior [19, 20]. Numerous researchers have investigated perceived obstacles to environmental behavior change. For instance, Blake [21] identified three primary barriers: intrapersonal factors (e.g., attitude and temperament), a lack of felt responsibility, and practicality concerns (e.g., perceived shortages of time, money, or information) [21].

In a societal study involving 151 residents of British Columbia by Gifford [22], participants were surveyed regarding the psychological obstacles they face when attempting household energy-saving behavior [22]. The study assessed behaviors

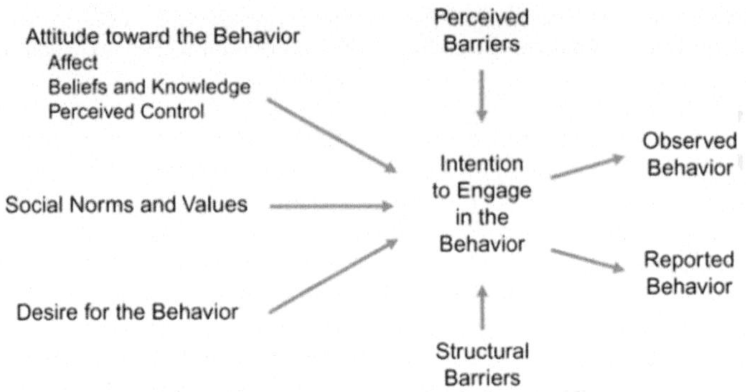

Fig. 6 Theory of behavioural choice (adapted from Gifford et al. [22])

using 11 items, such as "I turn off the television and computer when not in use." Five (out of the seven, as shown in Fig. 4) theorized barrier categories (Limited Cognition, Ideological Factors, Social Comparisons, Sunk Costs, and Discredence) were guided by their presumed relevance to the study's energy-related goals. The retained components in the principal component analysis were Denial, Conflicting Goals and Aspirations, Interpersonal Influences, Mission Impossible, Techno-salvation, and Ignorance (not knowing how to change). Ignorance was the only factor that did not show a significant correlation with the measured behaviors aimed at conserving energy which primarily consisted of low-cost actions. Authors proposed that the Ignorance barrier may be particularly relevant to difficult, high-cost behaviors.

Alagidede and Afego adopted a choice experiment to exploit variations in the intention of the young generation to engage in energy-efficient behaviours to identify barriers, and potential policy responses to support citizen action on climate change mitigation [13]. A choice experiment was conducted with students at the University of Ghana and the University of Nigeria to derive actionable insights on the demographic factors (e.g., gender, marital status, geographic location etc.), and psycho-social, cultural and structural factors (e.g., cognitive biases, desire, social norms and values, income, personality, beliefs and knowledge etc.), influence individuals' intentions to engage in sustainable energy use behaviours. In addition, the level of influence of these factors on the valuations of their willingness to pay for renewable energy options was studied with the help of the theory of behavioural choice [22] (Fig. 6).

Behaviour Changes for Sustainable Development

Consumer behavior change is shaped by various individual factors, including emotional influences and cognitive biases, as well as external factors like socio-economic disparities, cultural norms related to energy consumption, and fiscal policies such as carbon pricing [23]. Given that approximately 72% of global greenhouse gas emissions stem from household consumption, it is imperative to address household decision-making and behaviors alongside technological innovations to achieve decarbonization [2]. While technical energy efficiency solutions aim to reduce emissions per unit of production or consumption, they often face challenges like rebound effects, limiting their net impact on emissions reduction. On the other hand, behavioral solutions involve active efforts to change the nature or amount of consumption, including renouncement, reduction, substitution, and efficiency improvement.

Simulations conducted by Moberg et al. [23] suggest that a combination of technical and behavioral approaches could significantly reduce household carbon emissions [23]. A 2% annual reduction rate in technical energy efficiency alone might halve global emissions by 2050, but a combined approach with behavioral shifts could potentially reduce emissions by almost 76% (Fig. 5). Figure 6 also underscores the sheer magnitude of emissions reductions that behavioural change, in theory, can accomplish.

* The Paris Accord, also known as the Paris Agreement, is a landmark international treaty within the United Nations Framework Convention on Climate Change (UNFCCC). Adopted in 2015, the agreement aims to address climate change by limiting global warming to well below 2 °C above pre-industrial levels, with efforts to limit the increase to 1.5 °C (Figs. 7 and 8).

There exist many models pertaining to behavior and behavior change concerning pro-environmental actions. They can be classified as follows:

(a) Theories that attribute behavior to individual motivational factors,
(b) Contextual factors that elucidate behavior, and
(c) Group dynamics leading to enduring behavior modification.

In recent times, interventions have been implemented in the context of developing countries to enhance the effectiveness of sustainable development initiatives. These interventions aim to influence a range of behaviors, including water and energy usage,

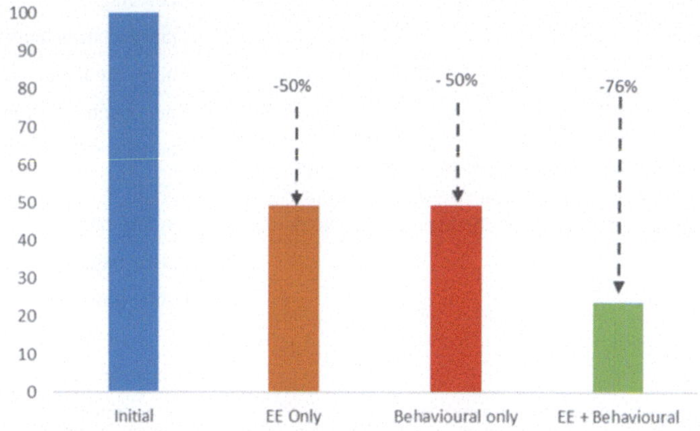

Fig. 7 Effect of a 2% annual reduction of GHG emissions due to energy efficiency only (orange), behavioral only (Red) and cumulated (Green), between 2015 and 2050 [2]

Fig. 8 Total gigatons of carbon displaced by behavioral change (in green) compared to INDC pledges (orange) and low-carbon infrastructure (blue) [2]

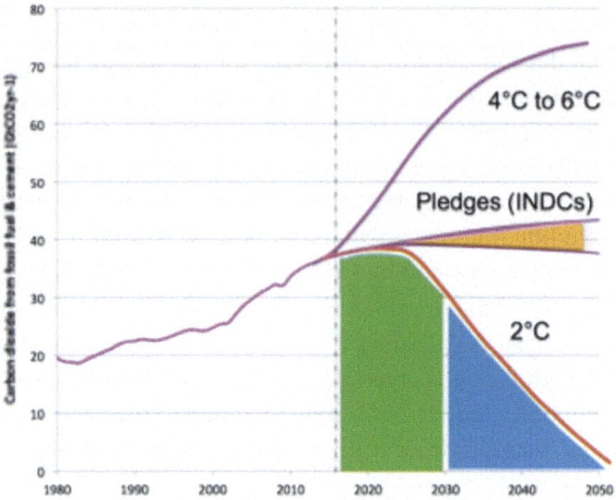

environmentally mindful purchases, waste reduction, and transportation (Table 1) [24].

Table 2 presents some of the successful interventions and one could notice that a 13.5% reduction in energy use could be achieved in offices through interventions such as prompting and competitions.

In the realm of household decision-making, literatures indicate a recurring trend where individuals tend to underestimate the potential energy savings achievable through the adoption of efficient or clean technologies. This phenomenon contributes

Table 1 Intervention tools and empirical applications [24]

Intervention tool	Case example
Informational	
Prompts	Recycling
Commitment	Transportation habits
Goal setting	Energy savings
Social model	Energy conservation
Feedback	Energy conservation
Structural	
Change in physical, technical, or organizational systems	Cycling rates
Legislation	Plastic bags
Price mechanisms	Public transport
Nudges	
Default settings	Green electricity
Simplification and framing of information	Food choice
Changes in physical environment	Food waste
Eliciting social norms	Hotel towel waste

Table 2 Examples of successful behaviour change interventions [24]

Country	Behaviors targeted	Intervention tools used	Results
Costa Rica	Household water consumption	Goal-setting; prompts, social norms	3.7–5.6% reduction in monthly water consumption
Denmark	Mobile phone purchases	Nudging	20% point increase in mobile phone repair; 7 × increase in purchase of second hand mobile phone
Norway, Switzerland, Denmark	Smart grid technology uptake	Default settings	2.5 × more likely to accept Smart grid installation in the opt-out condition
Kenya	Water purification	Nudges	Uptake rates rose from 10 to 60%
India	Daily commuting	Incentives	13% point increase in commuters traveling before peak times
Japan	Sustainable transportation	Feedback, Goal-setting	7.5% reduction in car use;68.6% increase in public transportation use
United States	Recycling	Commitment, feedback	25.4–40% increase in paper recycling
South Africa	Office energy efficiency	Prompts; competition	13.5% reduction in energy use
Denmark	Vegetable purchases	Nudges	61.3% increase in sales of pre-cut vegetables

to what is described as "the low adoption of energy-efficient technologies despite potentially large savings" [25]. A study by Norzalina et al. [26] analyzed consumers' intentions regarding the purchase of energy-efficient products [26]. The results revealed that consumers' perceived value of energy-efficient products is a pivotal factor influencing purchasing behavior. This implies that when consumers perceive energy-efficient products as superior in terms of energy and cost savings, as well as contributing to community and environmental benefits, their inclination to make a purchase is heightened.

Agencies and decision makers though give a thought on giving price-based incentives, it is understood that behavioral nudges are more cost-effective to implement compared to price-based incentives. The adoption behaviors of end-users are most probably influenced by informal institutions, traditions, and past habits, standard price-based incentives may not always achieve their intended outcomes or may necessitate complimentary non-price interventions [27].

Current research in behavioral economics proposes the concept of "nudging," which involves subtly influencing by educating producers and consumers by providing them with information on energy costs, environmental externalities, and other pertinent factors such as peer comparison of energy consumption. Altering intrinsic preferences can act as a potent mechanism for lowering energy demand. Nevertheless, given that individual decisions are made within social frameworks, elements such as institutional structures, social customs and norms, familial and kinship bonds, larger societal tensions, and governmental interventions can influence how individuals make decisions. In this context, one frequently discussed nudge involves employing social norms, specifically through peer comparisons. Experiments conducted by sending home energy reports including peer comparisons showed that, with repeated interventions, consumers have invested in energy-efficient appliances, as well as new habits adopted by consumers and integrated into their decision-making processes [3]. For example, Midden et al. [28] found that solely providing information is ineffective, and comparative feedback is more successful in reducing electricity consumption compared to individual feedback [28].

Psychologically, a significant limitation in the efficacy of nudging as a behavior change tool is its failure to address the attitudes, values, and beliefs that form the underlying motivations driving individuals to take action [29].

These insights highlight the need of policymakers and researchers prioritize stimulating sustainable behaviors and lifestyles, recognizing the substantial potential of behavioral change in achieving significant emissions reductions compared to low-carbon infrastructural supply or existing international agreements like the Paris Accord*. Addressing the nexus of demand-side options, lifestyles, barriers, and necessary behavioral changes is crucial for effective decarbonization [23].

Conclusions

Indeed, the recognition of the need for sustainable behavior change is crucial in addressing climate change. While policy measures, technological innovations, and systemic changes are essential components of climate action, individual and collective behaviors also play a significant role. Achieving the necessary reductions in greenhouse gas emissions and transitioning to a sustainable future requires shifts in lifestyles, consumption patterns, and everyday choices.

In the energy sector, the adoption of energy-efficient technologies are crucial practical considerations underlying climate change mitigation and adaptation efforts. However, despite evidence linking climate change to energy consumption behaviors, there is still much unknown about: (a) the behavioral and structural processes governing individuals' preferences for energy consumption, (b) the factors determining their relative importance, (c) how they mutually influence each other, and (d) how the driving forces interact to facilitate actual adaptation behaviors by individuals. This is significant because failure to mitigate and adapt to climate change could have severe repercussions on the economic and social well-being of large segments of the population in the years to come.

These findings of many studies suggest policy implications, indicating that governments should actively engage in heightening public consciousness about sustainable development. Merely increasing awareness may not inherently lead to diminished energy usage, given that subjective factors also hold considerable importance. The subjective attitudes of households can notably impact decisions to invest and encourage the adoption of more energy-efficient products and advocating for pro-environmental choices.

Encouraging green consumption through incentives may be challenging due to high welfare costs, and designing a feasible mechanism can be complex. Mandating reduced energy use is also not a viable option for developing countries where bolstering domestic consumption is crucial for maintaining stable economic growth. Instead, by altering consumption patterns and shaping the preferences of individual households, it becomes feasible to contribute to sustainable development in the energy sector.

References

1. Sovacool, B K et al. "Sociotechnical agendas: Reviewing future directions for energy and climate research." *Energy Research & Social Science*, 70 (2020): 101617.
2. Matos, Stelvia, et al. "Innovation and climate change: A review and introduction to the special issue." Technovation (2022): 102612.
3. Bahinipati, C S, Sirohi, R A, & Rao, S S. "Technological Innovations, Behavioural Interventions, and Household Energy Conservation: Policy Insights and Lessons". *Ecology, Economy and Society–the INSEE Journal*, 5.1 (2022), 63–87.
4. Fujii, H and Managi, S, "Economic development and multiple air pollutant emissions from the industrial sector", *Environmental Science and Pollution Research*, 23.3 (2016): 2802-2812.
5. Liu, J, Zhang, D, Cai, J, Davenport, J, "Legal systems, national governance and renewable energy investment: evidence from around the world", *British Journal of Management* (2019), https://doi.org/10.1111/1467-8551.12377.
6. Li, Xiangyu, et al. "Awareness, energy consumption and pro-environmental choices of Chinese households." *Journal of Cleaner Production* 279 (2021): 123734.
7. Zhang, Junze, et al. "Mapping the complexity of the food-energy-water nexus from the lens of Sustainable Development Goals in China." *Resources, Conservation and Recycling* 183 (2022): 106357.
8. Steg L, Perlaviciute G and van der Werff E. "Understanding the human dimensions of a sustainable energy transition". *Frontiers in Psychology*, 6:805 (2015), https://doi.org/10.3389/fpsyg.2015.00805.
9. Wieczorek-Kosmala, Monika, Dorota Marquardt, and Jarosław Kurpanik. "Drivers of sustainable performance in European energy sector." *Energies* 14.21 (2021): 7055.
10. Biresselioglu ME, Demir MH, Demirbag Kaplan M, Solak B. "Individuals, collectives, and energy transition: analysing the motivators and barriers of European decarbonization". (2020). *Energy Research and Social Science*, 66:101493. https://doi.org/10.1016/j.erss.2020.101493.
11. Krumm, Alexandra, Diana Süsser, and Philipp Blechinger. "Modelling social aspects of the energy transition: What is the current representation of social factors in energy models?." *Energy* 239 (2022): 121706.

12. Yin Bai, Yong Liu. "An exploration of residents' low-carbon awareness and behavior in Tianjin, China". *Energy Policy*, 61 (2013): 1261-1270.
13. Alagidede, I P and Afego, P N. "Climate risk perception and energy use behaviour among young people: experimental evidence from Ghana and Nigeria". *Thematic research proposal*, South Africa.
14. Kohler J, et al. "Modelling sustainability transitions: an assessment of ap proaches and challenges". *Journal of Artificial Societies and Social Simulation* (2018), 21(1):8. https://doi.org/10.18564/jasss.3629.
15. Upham P, Oltra C, Boso A. Towards a cross-paradigmatic framework of the social acceptance of energy systems. *Energy Research and Social Science* 8:100e12 (2015). https://doi.org/10.1016/j.erss.2015.05.003.
16. Wüstenhagen R, Wolsink M, Bürer MJ. "Social acceptance of renewable energy innovation: an introduction to the concept". *Energy Policy* 35.5 (2007). 2683e91. https://doi.org/10.1016/j.enpol.2006.12.001.
17. Devine-Wright P, Batel S, Aas O, Sovacool B, Labelle MC, Ruud A. "A conceptual framework for understanding the social acceptance of energy infrastructure: insights from energy storage". *Energy Policy* 107:27e31. (2017). https://doi.org/10.1016/j.enpol.2017.04.020.
18. Steinberger, Julia K., William F. Lamb, and Marco Sakai. "Your money or your life? The carbon-development paradox." *Environmental Research Letters* 15.4 (2020): 044016.
19. Chaiken, S., Stangor, C., 1987. Attitudes and attitude change. *Annual Review of Psychology* 38, 575–630.
20. Finger, M., 1994. From knowledge to action? Exploring the relationships between experiences, learning and behavior. *Journal of Social Issues* 50, 141–160.
21. Blake, J. "Overcoming the value-action gap in environmental policy: Tensions between national policy and local experience". *Local Environment* 4 (1999), 257–278
22. Gifford, R., Lacroix, K. and Chen, A., 2018. "Understanding responses to climate change: Psychological barriers to mitigation and a new theory of behavioral choice". *In Psychology and climate change* (2018): 161–183. Academic Press.
23. Moberg, K R et al., "Barriers, emotions, and motivational levers for lifestyle transformation in Norwegian household decarbonization pathways". *Climatic Change* 165.1(2021): 1–25.
24. Klaniecki, Kathleen, Katharina Wuropulos, and Caroline Persson Hager. "Behaviour change for sustainable development." *Encyclopedia of Sustainability in Higher Education; Leal Filho, W., Ed* (2018): 1–10.
25. Allcott, Hunt, and Dmitry Taubinsky. 2013. "The Lightbulb Paradox: Evidence from Two Randomized Experiments" NBER Working Paper No. 19713. https://doi.org/10.3386/w19713.
26. Zainudin, Norzalina, Jasmine Leby Lau, and Chandramalar Munusami. "Modelling household behavioural changes as an opportunity for sustainable home energy." *Environmental Economics and Policy Studies* 24.1 (2022): 73–97.
27. Thaler, Richard H and Cass R. Sunstein. 2009. Nudge: Improving Decisions about Health, Wealth, and Happiness. Penguin Books.
28. Midden, Cees JH, Joanne F Meter, Mieneke H Weenig, and Henk JA Zieverink. "Using Feedback, Reinforcement and Information to Reduce Energy Consumption in Households: A Field-experiment." *Journal of Economic Psychology* 3.1 (1983): 65–86. https://doi.org/10.1016/0167-4870(83)90058-2.
29. Newell P, Twena M, Daley F. Scaling behaviour change for a 1.5 degree world: challenges and opportunities. *Global Sustainability* 4, e22 (2021),1–13. https://doi.org/10.1017/sus.2021.23

Transforming Buildings Emission-Free for Sustainable Development: A Case Study

Sooraj K. Prabha

Abstract The rising population and its impact on the environment are big challenges that need to be addressed as a first priority. The rising population and activities for economic growth are causing threats to the ecosystem of this planet. The United Nations has envisaged policies for sustainable development and the goals set for the health of the planet are better known as Sustainable Development Goals. This chapter discusses the need of reducing carbon emission in every activity and the methods to reduce the emission. The building sector takes fourth position while considering global sector wise emissions. A case study is presented in the later part of the chapter to assess the carbon emission by buildings due to energy consumption. The possible/available methods to mitigate emissions thereby making the building emission-free are also discussed.

Keywords Sustainable development goals · Carbon emissions · Emission-free buildings

The rising population of mankind and depletion of natural resources is one of the biggest challenges that mankind is facing. According to the UN, around 1.6 billion people are homeless or live in inadequate living conditions. Providing shelter to such a huge mass is a challenging task [1]. The number of people living in extreme poverty conditions (less than \$1.9 per day, in 2022 the World Bank has revised the numbers to less than \$2.15 per day) was dropping at a rate of less than half a percentage a year till 2017. Indeed, the Covid pandemic has caused the poverty line to take a jump in reverse direction, making the situation more vulnerable (Fig. 1) [2]. A rise in extreme poverty indicates that the number of people who have limited access to economic resources has also increased.

The increasing population has a detrimental effect on the environment on many fronts. To mitigate the impact of rising population on the environment a sustainable

S. K. Prabha (✉)
Department of Mechanical Engineering, Vidya Academy of Science and Technology, Thrissur, Kerala, India
e-mail: sooraj.kp@vidyaacademy.ac.in

K. J. Sreekanth (ed.), *Energy Efficiency Improvements with Emission Abatement for Energy Sustainability*, https://doi.org/10.1007/978-981-97-9480-5_7

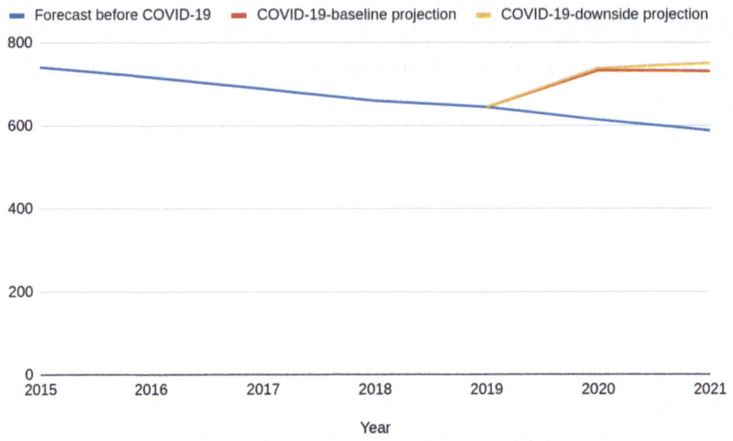

Fig. 1 Number of people living below $1.90 a day forecast before and after COVID-19 (millions)

development mechanism has to be implemented in every facet of life. The UN has envisaged sustainable development goals (SDG) for the prosperity of people and the planet. The rising population definitely increases the demand for energy and hence increases in energy production.

The increase in people living in extreme poverty conditions will also affect the goal of providing clean energy to all. Earlier, the UN had illustrated that the supply of electricity is a means of achieving development as it improves the economic status [3]. Accordingly, access to modern energy services will extend income generating activities which in turn reduce poverty [4]. Therefore, we can say that the UN's mission of removing poverty (SDG1) and providing clean energy to all (SGD7) go hand-in-hand, as getting energy to all improves economic activity. The clean energy production and the effective use of energy will definitely dampen the adverse effect of energy production on the environment. The sustainable development mechanisms on different fronts ultimately point towards the carbon emission reduction in all activities.

Low Carbon Emissions for Sustainable Development

Fast growing cities are the key contributors to carbon emission. It is essential to have a low carbon emissions society to achieve the goal of sustainable development [5]. There will be a sudden increase in the demand for energy in fast growing cities. The gap between ambitious targets of prevailing policies and the clean energy availability targets (SDG7) are becoming wider. This shows the importance of energy saving and its efficient use. According to the International Energy Agency (IEA), energy efficiency, electrification and behavioral change will provide 80% of the emissions reductions in the buildings sector by 2030 [6].

Carbon Emission

Carbon emission is broadly classified into two, artificial and natural emissions. The volcanic eruptions, animal and plant respirations, natural fires, etc. are the examples of natural emissions whereas artificial (manmade) include sources like industries, automobiles, power plants, etc. Whenever we discuss man made carbon emissions, emissions from factories/industries, automobiles, power plants, etc. come to our mind. Clearly these are the direct forms of carbon emission to the atmosphere. However, there are indirect forms of emission that contribute significantly to the total emission numbers, carbon emission from buildings is one of these kinds. To mitigate the indirect emissions from various sources it is important to estimate the emission equivalent in indirect forms. The estimation of emissions from the direct emission sources is much easier and direct when compared to the indirect emissions. In this chapter, carbon emission reduction strategies for sustainable development are presented with an emphasis on the building sector.

Global Sector Wise CO_2 Emissions

Figure 2 shows the distribution of CO_2 emissions in 2022 among various sectors. The energy and heat production has always been at the top in the list for decades. The building sector takes the fourth spot after energy production, transport and manufacturing (Fig. 2) [7]. Among buildings commercial buildings are the major contributors of carbon emission [8]. However, the total emission figures for the buildings sector is in the range of 8–10 in the last decade.

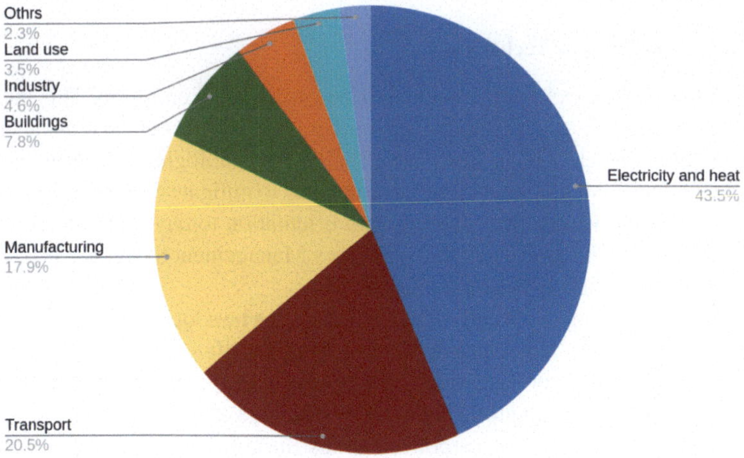

Fig. 2 CO_2 emissions by sector worldwide

Carbon Emission Reduction Strategies

The carbon emission reduction strategy is discussed in this section with emphasis on the emission reduction in buildings. The important strategies include, but not limited to, the following:

- enforcing standards and or policies,
- adopting low carbon technology,
- restricting energy utilization,
- adopting carbon neutrality measures, and
- education and awareness.

Enforcing Standards and Policies

This is one of the straight methods to reduce emissions, especially in buildings. This will have a sudden impact on the mitigation of CO_2 emission. This is successful and highly effective only among civilized and educated groups of people who respect the laws of the land. However, when sustainability norms are enforced to new buildings it will have a great impact on carbon emission. Needless to mention, this strategy of enforcing the standards and policies will have to be revised periodically.

In general, policies for sustainable development should come from governing bodies. Funding for research and development, modernization of grids, the introduction of smart grids, renewable energy targets, etc. usually come from governing bodies. Segregating the buildings with ratings and giving attractive incentives for top rated buildings also accelerate the growth towards sustainable development. To this end, carbon taxes proportional to the carbon footprint and emission trading schemes are other options in front of policy makers [9]. Governing quality, orientation of the government, and resources have a combined role to play in the path towards a clean energy transition [10].

Adopting Low Carbon Technology

Adopting low carbon technology is another way to mitigate emission. This includes material selection, use of alternate power sources, using cleaner energy, using energy efficient equipment, etc. Using sustainable building technologies in heating, ventilation and air conditioning systems is yet another way to mitigate carbon emission. The wise use of natural ventilation, shading, and orientation to daylight can reduce the need for mechanical heating and cooling. Energy Management Systems can optimize the use of lighting, air conditioners, heaters, etc.

For existing buildings, low carbon technologies can be adopted while doing maintenance or renovation. This can slightly increase the life cycle cost of buildings, however, it is inevitable for sustainable development. Proper use of materials during replacement improves the natural lighting in buildings and also reduces air conditioning loads. The use of low carbon materials during renovation and retrofitting can help to reduce emissions. For example, the use of recycled steel, bamboo, etc. reduces the carbon footprint of the building materials.

Restricting Energy Utilization

Strict adherence to good practices to be followed for low emissions is necessary. Changing lifestyles to reduce emission is crucial when it comes to large masses. Use of public transport, telecommuting, carpooling, etc. will reduce emissions by large numbers.

Restricting energy usage in high consumption areas will also improve the total emission numbers. This includes restricting the use of high power consuming types of equipment, and minimizing the use of low energy efficient equipment. To name a few, closing the doors and windows in air-conditioned rooms, switching off lights and other accessories while not in use, closing down used areas, etc. are some good practices.

Adopting Carbon Neutrality Measures

Carbon neutrality is the process of nullifying the effects of carbon emissions by removing an equivalent amount of carbon or by offsetting the emissions. This can be achieved by various means. The use of renewable or clean energy sources is one of them. Afforestation or planting trees to offset the emission is another method for achieving neutrality. Use of energy efficient lighting, use of energy efficient machines, use of energy efficient materials, etc. will reduce the emission. Reducing emission by taking such measures is also a strategy to achieve neutrality.

Carbon capture and storage is yet another option to achieve neutrality. The carbon capture and storage technologies are in the early stages of their development. Direct Air Capture system is one among them [11]. Apart from this there are other forms of carbon capture systems classified under Negative Emission Technologies (NETs) [12]. Coastal blue carbon, Terrestrial carbon removal and sequestration, Bioenergy with carbon capture and sequestration (BECCS), Direct air capture Carbon mineralization, and Geologic sequestration are considered as Negative Emission Technologies. These systems will have a significant role in mitigation carbon emission and climate change.

Education and Awareness

The efficacy of measures taken for sustainability highly depends on the end users. Irrespective of policies and norms, awareness among the users has a significant role to play in this regard. Conducting awareness programs and revising educational syllabi for sustainability are indirect means of achieving the target. Irrespective of the branch and discipline of study, the educational system must provide sufficient knowledge about the causes of climate change and its impact on the environment. It helps young minds to understand the connection between human activities and carbon emissions. Sustainable practices like energy conservation, waste reduction, etc. must be taught very early in schools to save the resources for the next generation. To address social, economic, cultural and environmental problems of the twenty-first century the UN is trying to incorporate all practices of sustainable development mechanisms into all aspects of education and learning through United Nations Decade for Education for Sustainable Development [13].

Carbon Emission: A Case Study

As mentioned above, the emissions by buildings contribute around 8–10% of total emissions globally. The Life-Cycle Carbon Emission Assessment can be performed in four life-cycle stages; product stage, construction stage, operational stage, and end-use stage [14]. For buildings, the emission during the operational stage is important as it lasts for decades, and it is the major contributor to total emissions in the life cycle. The emissions calculated here are in the operational stage of a building due to energy consumption.

In this section, a case study is presented that discusses carbon emission and carbon neutrality. The process discussed here can be replicated in any other sector with the same basic principles of operation. To this end, a portion of the academic block of an educational institution (Vidya Academy of Science and Technology, Kerala, India) is selected. The institute is situated in the southern part of India, in the state of Kerala. The estimation of carbon emission and exploring methods to neutralize the emissions are the key objectives of the study. The outcomes of the study can lead to saving energy, which is essential for sustainability. The first step towards saving energy is accounting the energy consumption with minute details. The process of accounting usage of energy, and its mapping is called energy audit.

Energy Audit

Energy auditing is a well-established process that needs no elaboration. However, for completeness, it is briefly explained here. The data collection for energy auditing is performed in two different stages i.e., preliminary energy auditing and detailed energy auditing. The preliminary audit is a preparatory phase wherein site visits, identifying different power-consuming equipment, number of occupants, etc. are determined. The actual energy auditing takes place in the detailed energy auditing phase.

The energy auditing is performed with the help of proper energy meters and power ratings of the equipment. With this, the average power consumption of equipment and accessories are found out. Once the average power consumption is known, the power consumption of each equipment over a period of time can be calculated. These power consumption numbers are fine-tuned with the data available in permanent energy meters installed locally in a room or floor. Once the average power consumption for every equipment is known, then, the energy usage of each area is identified. Further, the power consumption by each type of equipment is recognized by the same data. This information is essential for taking remedial actions to meet the targets.

The power consumption per year for a floor of the academic block is depicted in Fig. 3. Clearly, the power consuming intensities are identified for the floor area, and similar mapping is done for all the floors of the building under consideration. This is essential for implementing energy saving measures. Similarly, the energy

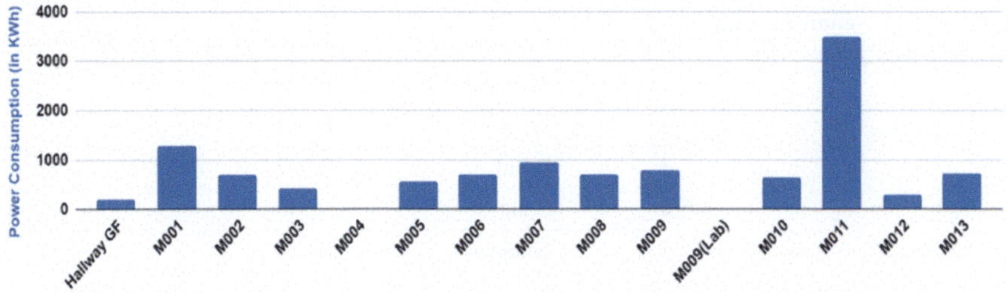

Fig. 3 Yearly power consumption distribution of ground floor of academic block (in kWh)

Fig. 4 Yearly consumption pattern for various equipment

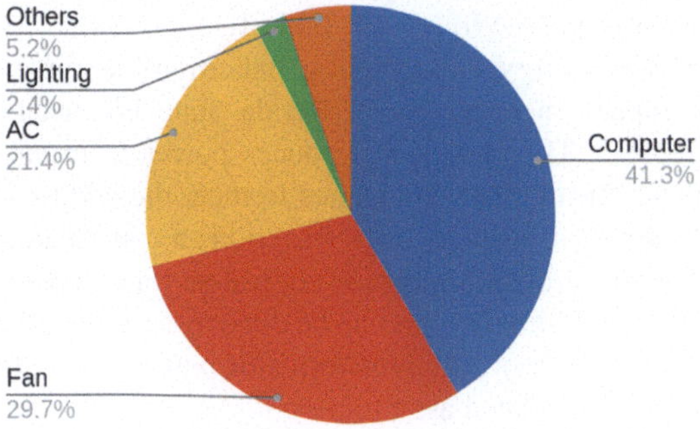

consumption mapping is done for each type of equipment installed in the building. For example, the consumption pattern for various equipment is as given in Fig. 4.

From Figs. 3 and 4, the major power-consuming equipment identified is computers, which are installed in Room M011 (Fig. 3). The next major power consumption is by fans and air conditioners installed for cooling; followed by lightings. All other equipment put together contributes around 5% of total power consumption. This type of mapping can be performed for buildings with large and small areas.

Carbon Emission Per Unit Power Consumed

The emission equivalent is a measure used to compare emissions from various sources based on their global warming potential. The power generation sector is a primary contributor to total carbon emission (Fig. 2). There are various forms of power sources like fossil fuel, renewable energy and nuclear energy. The renewable energy sources produce power with a lesser carbon fruit print on the environment. On the other hand, coal fired power plants are top in the list. The power production and distribution companies may produce and distribute power from various sources through the same grid. Therefore, the power distribution over a period of time is a blend of power from various sources like coal, wind, solar and nuclear. Consequently, the calculation of

Table 1 CO_2– equivalent for various power plants [15]

Sl. no	Power plant	CO_2 equivalent (kgs/kWh)
1	Thermal (Coal)	0.820
2	Hydro	0.024
3	Wind	0.012
4	Solar	0.048
5	Nuclear	0.012

the carbon equivalent of unit power becomes complicated. Each primary source has its respective CO_2– equivalent. The carbon equivalent for different power sources are given in the following Table 1.

The energy consumption pattern for a typical power generation distribution company in South India, Kerala State Electricity Board (KSEB) is as shown in Table 2. The company produces power from various sources and also purchases power from other companies to meet the additional requirements. In this case, the power distributed through the grid is a combination of power from various sources. Therefore, carbon emission equivalent for unit power generated will have to be calculated accordingly. This can be done by assuming that the power distributed is a linear combination of different forms of power. Therefore, the average emission in each year is calculated as,

$$Average\ Emission\ (per\ kWh) = \sum_{i}^{n} (EF_i * P_i)$$

Table 2 The power produced/distributed by the state owned KSEB from various sources are given in percentage

Year	Hydro	Wind	Thermal	Solar	Nuclear	Emission
2010–11	44.5	0.4	46.3	0	8.8	0.391
2011–12	48.1	0.4	48.9	0	2.6	0.413
2012–13	32.8	0.5	63.8	0	2.9	0.531
2013–14	43.8	0.4	52.6	0	3.2	0.442
2014–15	87	0.8	12.1	0	0	0.120
2015–16	95.2	0.8	3.9	0	0	0.055
2016–17	36.3	0.8	62.8	0.1	0	0.524
2017–18	29.6	0.6	60.3	0.4	9.1	0.503
2018–19	37.5	0.5	55	0.5	6.4	0.461
2019–20	29.1	0.6	59.8	0.5	9.8	0.499
2020–21	34.9	0.5	53.7	0.8	9.9	0.450

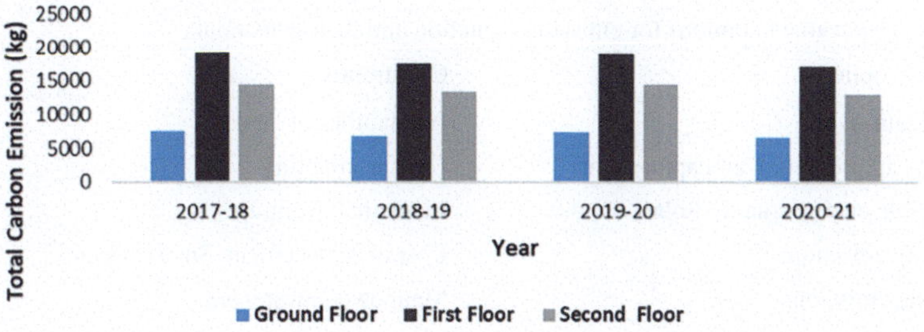

Fig. 5 The carbon emission numbers by different floors of the building under study for four years

where P_i is the fractional power from each source and EF_i is the respective emission factor for each source. The fractional power generated/distributed in each year is also given in the table and their respective emission factor is also given. From Table 2, it is observed that when the output from hydro-electric power plants is maximum the emission is less, and when the contribution from thermal power is high the emission is high.

Carbon Emission Estimation

Figure 5 shows floor-wise CO_2 emissions of the building under study. Since the first floor accounts for the majority of electricity consumption, the first floor emits a higher quantity of emissions. Even if the power consumption pattern remains the same over the years, the carbon emission pattern can vary depending on the blend of unit power consumed. This factor is overlooked in many of the studies reported on carbon emission. Thus, the total power consumption is found to be 81.5 MWh for a particular year. This is equivalent to 41.16 tonnes of CO_2 generated.

Carbon Neutrality

Once the emission numbers are estimated the next step is to take measures to reduce emission, and thereby zeroing in achieving neutrality. There are various strategies to reduce CO_2 emission caused by energy usage. This is done by weighing the available options by considering constraints for each. Some of the available options at the reach and their constraints are given in Table 3. In general, they may not be limited to the following ones given in the table.

Among the renewable energy sources, solar power is comparatively easy to install. The output power of solar power plants can be easily customized as per our needs, ranging from the smallest to the largest plants. If there is no additional land available for the installation of solar panels, the terrace of the existing building can be used

Table 3 The available options for emission reduction and their constraints

Available options	Constraints
Afforestation	Availability of land
Carbon capture/Direct air capture unit	Cost, availability
Renewable energy source (Solar)	Land area, Built up space
Reducing emissions	Cost of replacement, Smart systems
Avoiding emissions	Manpower, awareness

for the purpose. The terrace of the building under study is around 150 m in length and 20 m in width, which gives an area of 3,000 m^2. This area can be utilized to install solar power plants that can offset approximately 19–22% of total emissions. A Direct Air Capture system that can also be installed on the same terrace, will reduce emissions by another 20%.

Planting trees that can grow in the local climate conditions is another viable option to offset emissions. If half a hectare land is available, another 6–8% emission can be reduced by offsetting emissions by carbon sequestration. For biological carbon sequestration by planting trees, the native species seen abundantly like Cocos nucifera (Coconut tall), Artocarpus heterophyllus (Jackfruit), Mangifera indica (Mango), Azadirachta indica (Neem), Moringa oleifera (Drumstick tree), Phyllanthus emblica (Indian Gooseberry) Cassia fistula (Indian Laburnum), and Ficus religiosa (Sacred Fig), etc. can be considered. As the majority of these yield fruits, there is an additional benefit of income from afforestation. Planting trees alongside existing agricultural land so as to get environmental and economic benefit is called agroforestry which is also an option that can be considered.

It is observed that replacing obsolete equipment with energy efficient ones [with the same capacity] will reduce the energy consumption by a good margin. For example, replacing lights with energy efficient LEDs, and the use of 5star rated electrical equipment and accessories can reduce power consumption considerably. (In India, Star rating is a label given to segregate the electrical equipment in terms of energy efficiency with 5stars being the most energy-efficient.) It is estimated that 41–43% reduction in emission can be achieved with this activity for the case study considered here. The remaining part will be eliminated by avoiding emissions. Therefore, to make the building carbon neutral for the operating stage, the proportion of reduction can be as shown in Fig. 6.

Conclusions

The efficient use of energy can reduce a significant amount of emission in almost all sectors. Needless to mention that saved energy is equivalent to energy produced by the same amount. In addition to reduction in emission, there are additional benefits of saving energy. Saving energy helps to have financial security and energy security.

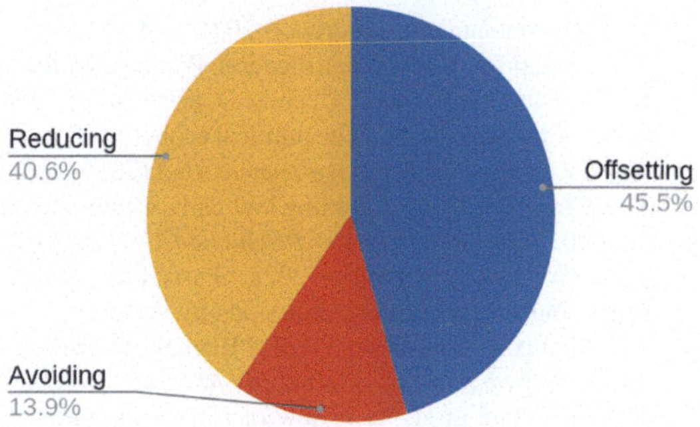

Fig. 6 Suggested neutrality measures for the case study

The measures taken to save energy are often beneficial for protecting health too. The saved energy can be utilized to meet the basic needs of needy factions of society without adding additional infrastructure. Further, this is a step closer to providing affordable and sustainable energy to all, which is one of the sustainable development goals envisaged by the UN (SDG7).

Summary

The increase in population and their basic demands are the biggest challenges to be addressed. The increase in extreme poverty makes the situation more difficult to address. The United Nations has envisaged sustainable development goals to meet the basic requirements of the people and for the wellbeing of the planet. The sustainable development mechanisms necessitates the need of carbon emission reduction in all sectors. When we analyze the carbon emission figures among different sectors the buildings section takes the fourth spot.

Reducing emissions by reducing energy consumption is one of the direct ways of mitigating emission in buildings. The methods to reduce emissions are discussed in detail with an emphasis of reduction in buildings.

In this chapter a case study is presented to calculate the emission in buildings due to energy consumption and the methods to mitigate it. The academic block of an educational institution was taken for the study. The energy expenditure mapping was done by energy auditing through different phases. The available options to reduce consumption are examined and the possibility of making the building carbon neutral is investigated. The emission is found to neutralize through reducing emission, offsetting emission and avoiding emission.

References

1. https://www.un.org/development/desa/dspd/2020/03/resolution-homelessness.

2. https://unstats.un.org/sdgs/report/2021/goal-01/.
3. United Nations. Rural Electrification. Prepared by the secretariat of the economic commission for Asia and the Far East. Geneva: Department of Economic Affairs; 1954.
4. Sovacool, Benjamin K. "The political economy of energy poverty: A review of key challenges." *Energy for Sustainable Development* 16.3 (2012): 272-282.
5. Lee, Chew Tin, et al. "Enabling low-carbon emissions for sustainable development in Asia and beyond." *Journal of Cleaner Production* 176 (2018): 726–735.
6. IEA. "World energy outlook 2022." Paris, France: International Energy Agency, 2022.
7. https://ourworldindata.org/emissions-by-sector.
8. Lu, Mengxue, and Joseph Lai. "Review on carbon emissions of commercial buildings." *Renewable and Sustainable Energy Reviews* 119 (2020): 109545.
9. Sizirici, Banu, et al. "A review of carbon footprint reduction in construction industry, from design to operation." *Materials* 14.20 (2021): 6094.
10. Arndt, Channing, et al. *The political economy of clean energy transitions.* Oxford University Press, 2017.
11. Browne, Stephen. *Sustainable development goals and UN goal-setting.* Taylor & Francis, 2017. Climate Intervention: Carbon Dioxide Removal and Reliable Sequestration (2015) Chapter: 2 Carbon Dioxide Removal.
12. Board, Ocean Studies, and National Academies of Sciences, Engineering, and Medicine. "Negative emissions technologies and reliable sequestration: a research agenda." (2019).
13. Buckler, C., and H. Creech. "UN Decade of Education for Sustainable Development (2005–2014) FINAL REPORT SUMMARY Shaping the Future We Want UN Decade of Education for Sustainable Development." *United Nations Educational, Scientific and Cultural Organization: Paris, France* (2014).
14. Fenner, Andriel Evandro, et al. "The carbon footprint of buildings: A review of methodologies and applications." *Renewable and Sustainable Energy Reviews* 94 (2018): 1142–1152.
15. Schlömer, Steffen, et al. "Annex III: Technology-specific cost and performance parameters." *Climate change 2014: Mitigation of climate change: Contribution of working group III to the fifth assessment report of the Intergovernmental Panel on Climate Change.* Cambridge University Press, 2014. 1329-1356.

Energy Policy Analyses, Energy Transition and Sustainability

K. J. Sreekanth

Abstract The global energy landscape stands at a pivotal crossroads, demanding a paradigm shift towards sustainability to address environmental concerns, resource scarcity, and geopolitical uncertainties. Effective energy policy analyses, a well-managed energy transition, and a commitment to sustainability are essential components for creating a resilient and environmentally responsible energy future. Policymakers, researchers, scientists, businesses, and society must work together to address challenges and promote the adoption of sustainable energy practices. In the pursuit of effective energy governance, policy analyses emerge as the cornerstone. This aspect involves the comprehensive evaluation of policies governing energy production, distribution, and consumption. A nuanced understanding of the intricate web of regulations, incentives, and market dynamics is essential to navigate the complexities of the energy sector. Robust policy analyses provide decision-makers with invaluable insights into the efficiency, effectiveness, and socio-economic impact of existing policies. Moreover, they serve as a compass for crafting future policies that align with broader societal goals, including environmental sustainability, economic development, and energy security. Central to the discourse on sustainable energy is the concept of energy transition—a deliberate shift from conventional, often carbon-intensive, energy sources to cleaner, renewable alternatives. This transition necessitates a multifaceted approach, encompassing technological innovation, infrastructure development, and policy incentives. Renewable energy sources, such as solar, wind, and hydropower, play a pivotal role in diversifying the energy mix, reducing greenhouse gas emissions, and enhancing energy security. Energy transition not only mitigates the adverse environmental impact of traditional energy sources but also fosters economic opportunities through the growth of green industries and job creation. The overarching goal of energy transition and policy analyses is to achieve sustainability—a delicate equilibrium between meeting current energy needs and safeguarding the resources for future generations. Sustainability in the energy context goes beyond environmental considerations; it encapsulates economic

K. J. Sreekanth (✉)
Energy and Building Research Center, Kuwait Institute for Scientific Research, Kuwait City, Kuwait
e-mail: sreekanthkj@gmail.com; sreekanthkj@kisr.edu.kw

K. J. Sreekanth (ed.), *Energy Efficiency Improvements with Emission Abatement for Energy Sustainability*, https://doi.org/10.1007/978-981-97-9480-5_8

viability and social equity. A sustainable energy framework ensures that development is not pursued at the expense of environmental degradation or social injustice. It calls for a holistic approach that integrates renewable energy deployment, energy efficiency measures, and the adoption of circular economy principles. This chapter underscores the urgency of concerted efforts to redefine our energy trajectory, ensuring a prosperous and sustainable future for generations to come by a harmonious integration of energy policy analyses, energy transition strategies, and sustainability principles. All must collaborate to navigate challenges, seize opportunities, and chart a course towards a resilient, low-carbon, and equitable energy landscape.

Keywords Energy landscape · Policy · Carbon economy · Resource availability · Energy sources · UN SDG

Introduction

Energy policy studies encompass a multidisciplinary field that delves into the formulation, implementation, and evaluation of policies governing the production, distribution, and consumption of energy resources. As societies navigate the challenges of the twenty-first century, the nexus between energy policy, energy transition, and sustainability has emerged as a critical focal point. This chapter aims to explore the intricate interplay between these elements, shedding light on the importance of well-crafted energy policies in steering nations towards a more sustainable and resilient future.

To understand the current landscape of energy policy studies, it is imperative to delve into the historical evolution of energy policies. The twentieth century witnessed a reliance on fossil fuels, driving economic growth but also ushering in environmental challenges. Policies were often crafted to ensure energy security and promote economic development, with environmental considerations taking a backseat [1, 2].

Energy Policy: A Foundation for Transition

Energy policy serves as the foundation upon which a nation's energy landscape is built. It encompasses a set of rules, regulations, and strategic objectives that guide the production, distribution, and consumption of energy resources. Effective energy policies set the trajectory for a nation's energy transition, influencing the types of energy sources deployed, infrastructure investments, and the overall resilience of the energy sector.

The historical evolution of energy policies reflects shifting priorities and responses to emerging challenges. Early policies often prioritized energy security and economic growth, leading to a heavy reliance on fossil fuels. However, as awareness of environmental impacts grew, contemporary energy policies are increasingly focused on sustainability, resilience, and the integration of renewable energy sources.

This comprehensive section examines energy policy studies in different countries, offering a comparative analysis of diverse approaches, challenges faced, and lessons learned. From industrialized nations to emerging economies, understanding the unique contexts within which energy policies are formulated provides valuable insights into the global energy landscape.

United States of America (USA)

The USA has undergone a significant transformation in its approach to energy policy over the decades. From the oil shocks of the 1970s to recent efforts to enhance energy security and promote sustainability, the historical evolution of U.S. energy policy is explored, key milestones, legislative acts, and shifts in priorities are analyzed to provide a comprehensive overview. The U.S. has experienced a boom in natural gas production, driven by advancements in shale gas extraction [3].

Examining the contemporary landscape, the challenges and opportunities facing U.S. energy policy is delved and topics include the role of shale gas and oil in the energy mix, the transition to renewable energy, and the impact of federal and state-level policies on the energy sector. Some states have implemented or are considering policies to reduce carbon emissions, such as Clean Energy Standards (CES) that mandate a certain share of electricity to come from low-carbon sources. The U.S. Environmental Protection Agency (EPA) plays a key role in implementing and enforcing regulations. Additionally, the role of technology and innovation in shaping the future of U.S. energy policy is explored [3].

China

China, as the world's largest energy consumer, has undergone rapid industrialization and urbanization, driving a surge in energy demand. How China has navigated the challenges of balancing economic growth with environmental sustainability and the role of coal, renewable energy deployment, and technological advancements in China's energy policy are scrutinized [4].

Given its global influence, China's energy policy decisions have far-reaching implications. The Belt and Road Initiative, renewable energy investments, and China's commitment to carbon neutrality are analyzed in the context of international cooperation and the impact on global energy dynamics. China has committed to peak its carbon emissions by 2030 and achieve carbon neutrality by 2060. Pilot programs for emissions trading have been launched in several provinces to create economic incentives for reducing carbon emissions. China has implemented policies to control the growth of coal consumption, aiming to reduce reliance on coal in favor of cleaner energy sources [4–6].

European Union (EU)

The European Union (EU) has been at the forefront of efforts to transition to a low-carbon economy. The EU's ambitious renewable energy targets, emissions reduction commitments, and the role of policy instruments such as the European Green Deal are examined and the effectiveness of regional collaboration and the challenges faced by individual member states in aligning their energy policies are discussed [7]. The EU has set ambitious targets for the share of renewable energy in its overall energy consumption, aiming to achieve at least 32% by 2030 [8].

Energy security is a perennial concern for the EU, given its dependence on external energy sources. How the EU navigates the delicate balance between ensuring a reliable energy supply and promoting sustainability is explored and geopolitical considerations, including relationships with energy-producing nations, are examined in the context of EU energy policy [7]. The Energy Performance of Buildings Directive (EPBD) sets standards for energy performance and efficiency in buildings [9]. The EU has introduced the European Green Deal, a comprehensive plan to make the EU's economy sustainable and achieve climate neutrality by 2050. The Emissions Trading System (ETS) is a cap-and-trade system designed to reduce greenhouse gas emissions in the industrial sector [10]. The Connecting Europe Facility (CEF) supports the development of trans-European networks in the energy sector, enhancing electricity and gas infrastructure [11].

India

India, with its vast population and diverse energy needs, faces unique challenges in crafting energy policies. How India addresses the dual goals of energy access for all and sustainable development and the role of renewable energy, rural electrification initiatives, and the integration of clean energy technologies into the energy mix are discussed [12]. As an emerging economy, India grapples with the need for economic development and poverty alleviation alongside its commitment to climate action. India's efforts to balance these priorities, examining policy measures, international collaborations, and the role of renewable energy in achieving sustainable development goals are analyzed [13].

India has set National Solar Mission, an ambitious target for solar power capacity installation, aiming to achieve a certain capacity [14]. India is a signatory to the Paris Agreement and has committed to reducing the emissions intensity of its GDP and increasing the share of non-fossil fuel-based energy capacity. The National Clean

Air Programme (NCAP) aims to address air pollution issues by promoting cleaner technologies and reducing reliance on fossil fuels.

The Deen Dayal Upadhyaya Gram Jyoti Yojana (DDUGJY) scheme focuses on strengthening rural electricity infrastructure and improving the reliability of power supply in rural areas [15]. The National Mission for Enhanced Energy Efficiency (NMEEE) promotes research and development in energy efficiency and the adoption of energy-efficient technologies [16].

Brazil

Brazil has been a global leader in the use of biofuels, particularly ethanol derived from sugarcane. Brazil's success in integrating biofuels into its energy matrix, examining policy measures, technological advancements, and the impact on energy security and environmental sustainability are explored [17]. The Proálcool Program encourages the production and use of ethanol as a fuel for transportation [18].

Despite its success in renewable energy, Brazil faces environmental challenges related to deforestation and the impact on ecosystems. Brazil addresses these challenges within the framework of its energy policies, highlighting the importance of balancing economic development with conservation efforts [17]. Brazil has significant offshore oil reserves in the pre-salt layer. Policies support the exploration and development of these reserves to enhance energy security and economic growth [19]. The National Plan for Energy Efficiency (PNEf) has implemented energy efficiency programs to improve energy consumption patterns across various sectors [20, 21]. Brazil has one operating nuclear power plant (Angra 1) and plans for further nuclear power development [22]. The Luz Para Todos (Light for All) program focuses on expanding electricity access to rural and remote areas, contributing to social development [23].

Japan

The Fukushima nuclear disaster in 2011 prompted Japan to reassess its energy policies, particularly its reliance on nuclear power. Japan has shifted its energy strategy, embracing renewables, energy efficiency measures, and a diversified energy mix beyond 2011. The role of public opinion, regulatory changes, and the impact on Japan's energy security are discussed [24].

Japan is known for its innovation in energy technologies, including advancements in smart grids and energy storage. Japan leverages technological solutions in its energy policies, fostering innovation and resilience in the face of evolving energy challenges [25]. The Feed-in Tariff system has been in place to encourage the development of renewable energy projects by providing favorable tariffs for electricity generated from renewable sources [26].

Kingdom of Saudi Arabia (KSA)

Saudi Arabia has set ambitious goals for the development of renewable energy, particularly solar and wind power. The country aims to generate a significant portion of its electricity from renewable sources by 2030. The Renewable Energy Project Development Office (REPDO) has been established to oversee the implementation of renewable energy projects and attract investments in the sector [27].

Vision 2030 outlines the broader economic and social transformation of Saudi Arabia, and energy is a crucial component of this vision. The plan includes increasing the share of renewables, improving energy efficiency, and optimizing the use of natural resources [28].

The NREP is a key initiative under Vision 2030, focusing on the development of renewable energy projects. It aims to contribute to the Kingdom's economic diversification and reduce its carbon footprint. The King Salman Renewable Energy Initiative, launched in 2019, is part of the NREP and aims to add 60 GW of renewable energy capacity by 2030 [29, 30].

Challenges and Best Practices

While each country has its unique set of circumstances and priorities, common challenges and themes emerge in the realm of energy policy. Drawing on the experiences of different countries, this section distills lessons learned and identifies best practices in energy policy formulation and implementation. Emphasizing the importance of flexibility, stakeholder engagement, and adaptive governance, the aim is to provide insights that can inform future energy policy decisions on a global scale. The major challenges and best practices are presented in Table 1.

Energy Transition: A Paradigm Shift

Energy transition refers to the gradual shift from conventional, often carbon-intensive, energy sources to cleaner, more sustainable alternatives. Drivers of energy transition include environmental concerns, technological advancements, and the economic viability of renewable energy sources. The urgency of mitigating climate change has accelerated the pace of this transition, compelling nations to reevaluate their energy portfolios [31].

Renewable energy sources, such as solar, wind, hydro, and geothermal, play a pivotal role in the energy transition. Policies that incentivize the development and

Table 1 Challenges and best practices

Features	Challenges	Best practices
Diverse energy sources	Balancing the mix of energy sources to ensure reliability, affordability, and sustainability	Conduct comprehensive assessments considering the country's resource availability, technological feasibility, and environmental impact
Political and regulatory stability	Achieving long-term policy continuity amid changing political landscapes	Develop bipartisan support and establish independent regulatory bodies to provide stability
Technology and innovation	Keeping up with rapidly evolving technologies and fostering innovation	Invest in research and development, collaborate with industry and academia, and regularly update policies to incorporate technological advancements
Energy security	Ensuring a secure and resilient energy supply	Diversify energy sources, invest in domestic resources, and establish strategic reserves
Environmental concerns	Mitigating environmental impacts and addressing climate change	Set clear emission reduction targets, promote renewable energy, and implement stringent environmental regulations
Economic viability	Balancing economic growth with sustainable energy practices	Implement policies that encourage investment in clean energy technologies, job creation, and economic diversification
Energy access and affordability	Ensuring universal access to affordable energy	Implement targeted programs for rural electrification, energy efficiency, and financial assistance for vulnerable populations
Infrastructure development	Building and upgrading infrastructure to accommodate new technologies and energy sources	Invest in modern, resilient infrastructure and promote public–private partnerships
Public awareness and participation	Garnering public support and awareness for energy policies	Conduct public consultations, education campaigns, and involve stakeholders in the policy-making process
International collaboration	Coordinating efforts with global energy and climate initiatives	Engage in international collaborations, share best practices, and participate in global agreements to address common challenges

integration of these sources into the energy mix are crucial for reducing carbon emissions and building a more sustainable energy infrastructure. The success of the transition is contingent upon the alignment of policies with technological advancements and market dynamics [31–33].

The traditional model of centralized power generation is being challenged by a shift toward more decentralized energy systems. This includes the integration of distributed renewable energy sources and advancements in energy storage technologies [34].

Improving energy efficiency is a key component of the transition. This involves adopting technologies and practices that reduce energy waste, enhance the efficiency of energy use in buildings, transportation, and industrial processes [35].

Many countries see the energy transition as an opportunity to diversify their economies. By investing in renewable energy technologies, nations aim to create new industries, generate employment, and reduce their dependence on a specific set of energy resources [36].

Governments play a crucial role in driving the energy transition through policy and regulatory frameworks. Incentives, subsidies, and regulations can influence the adoption of renewable energy and the overall direction of the energy market [37].

Companies and individuals are increasingly recognizing the importance of sustainability. Corporate social responsibility (CSR) initiatives often include commitments to reduce carbon footprints, adopt renewable energy, and contribute to broader environmental goals [38].

The energy transition is closely linked to global efforts to mitigate climate change. By reducing reliance on fossil fuels, countries aim to lower greenhouse gas emissions and limit the impacts of climate change [39].

Sustainability: The End Goal

Sustainability, in the context of energy, extends beyond environmental considerations to encompass economic, social, and geopolitical dimensions. A holistic approach to sustainability requires energy policies to address the triple bottom line—people, planet, and profit. Policies must foster economic growth, ensure social equity and inclusivity, and contribute to the global goal of environmental preservation [40].

At the core of sustainability is environmental stewardship. Energy policies must actively promote the reduction of greenhouse gas emissions, conservation of biodiversity, and the preservation of ecosystems. This involves setting ambitious emission reduction targets, investing in carbon capture technologies, and encouraging the adoption of sustainable practices across all sectors [41].

Social Equity and Inclusivity

Sustainability is intrinsically tied to social equity and inclusivity. Energy policies should prioritize universal energy access, especially in underserved communities. Inclusive policies that consider the socio-economic impact of energy transitions on vulnerable populations are essential to prevent exacerbating existing disparities [31, 42].

Economic Resilience

Economic sustainability requires energy policies to foster innovation, job creation, and resilience in the face of global challenges. A transition to sustainable energy sources can stimulate economic growth, create new employment opportunities, and enhance a nation's competitiveness in the global market [43].

Challenges and Trade-Offs

Economic Realities

Despite the overarching goals of sustainability, energy policies often encounter economic constraints and trade-offs. The transition to renewable energy may necessitate initial investments that strain national budgets. Policymakers must navigate the delicate balance between economic growth and the long-term benefits of sustainable practices [44].

Political Dynamics

Political resistance, both domestically and internationally, poses a significant challenge to the implementation of sustainable energy policies. Fossil fuel industries, entrenched interests, and geopolitical considerations can hinder the adoption of progressive policies. Policymakers must navigate these complexities to ensure the viability of sustainability-focused initiatives [31, 45].

Technological and Infrastructural Challenges

The integration of renewable energy sources into existing energy infrastructures poses technological and infrastructural challenges. Energy storage, grid modernization, and the development of new technologies require concerted efforts and strategic planning. Energy policies must facilitate research and development while ensuring a smooth transition to more sustainable energy systems [46, 47].

Learning from Global Experiences

Germany's Energiewende

Germany's Energiewende, or energy transition, serves as a notable case study. This ambitious policy initiative aims to shift the country towards a more sustainable energy system, emphasizing the importance of renewable sources, energy efficiency, and carbon neutrality. Analyzing Germany's successes and challenges provides valuable insights for other nations navigating similar transitions [48, 49].

The Nordic Model

Nordic countries, including Denmark and Sweden, have embraced renewable energy and energy efficiency as integral components of their energy policies. These nations provide examples of successful integration of renewable energy into the national grid, demonstrating the feasibility of ambitious sustainability goals [50].

The Indian Experience

India has set ambitious targets for increasing its renewable energy capacity. The National Solar Mission and the National Wind Mission are key components of the country's efforts to expand solar and wind energy production [51]. The country aims to achieve 175 GW of renewable energy capacity by 2022, with a significant portion coming from solar (100 GW) and wind (60 GW) [52].

India has made substantial investments in solar power projects, including large-scale solar parks and distributed solar installations. The government has implemented various policies and incentives to promote solar energy adoption. India has implemented various energy efficiency programs to reduce energy consumption in industries, buildings, and transportation. Initiatives such as the Perform, Achieve, and Trade (PAT) scheme target energy-intensive industries to improve their energy efficiency [53].

India has expressed interest in promoting electric vehicles as part of its efforts to reduce the environmental impact of the transportation sector. The Faster Adoption and Manufacturing of Hybrid and Electric Vehicles (FAME) scheme provides incentives for the adoption of EVs and charging infrastructure [54]. India has been working on extending access to electricity in rural areas. The Saubhagya scheme, launched in 2017, aims to provide electricity to all households, emphasizing the use of clean and sustainable energy sources [55]. India actively engages in international collaborations to address energy and environmental challenges. The International Solar Alliance (ISA), launched by India and France, aims to promote solar energy globally [56].

Challenges in Developing Nations

Examining the energy policies of developing nations reveals unique challenges. Balancing energy access, economic development, and environmental sustainability requires nuanced policies. Case studies from countries like India and Brazil illustrate the complexities and opportunities associated with sustainable energy transitions in diverse socio-economic contexts [57–59].

Some key challenges and best practices that developing countries may encounter during the examination of energy policies are presented herewith in Tables 2 and 3.

Future Prospects and Emerging Trends

Role of Technology

The future of energy policy will be intricately linked to technological advancements. Artificial intelligence, smart grids, energy storage solutions, and breakthroughs in clean energy technologies will shape the landscape of energy transition. Policies that encourage innovation and adaptability will be crucial for navigating the evolving technological frontier [60].

International Collaboration

Global challenges demand global solutions. The interconnected nature of energy systems necessitates international collaboration. Multilateral agreements, knowledge sharing, and coordinated efforts are essential for addressing climate change and ensuring a sustainable energy future. Policymakers must actively engage in diplomatic efforts to foster cooperation and shared responsibility [61, 62].

Decentralized Energy Systems

The rise of decentralized energy systems, including community-owned renewable projects and off-grid solutions, represents a paradigm shift in energy production and consumption. Policies that support decentralization empower local communities, enhance resilience, and contribute to broader sustainability goals [63].

The Link Between Energy Policy, Energy Transition, and Sustainability in Oil-Rich Countries

Oil-rich countries stand at a critical crossroads in the twenty-first century as they grapple with the intricate interplay between energy policy, the imperative for an energy transition, and the global call for sustainability. Historically reliant on fossil

Table 2 Challenges and impact

Constraints	Challenges	Impact
Limited financial resources	Insufficient financial resources may constrain the ability to invest in new and cleaner energy technologies	Limited funding can impede the development of renewable energy projects and the modernization of energy infrastructure
Infrastructure deficiencies	Outdated or inadequate energy infrastructure may limit the reliable supply and distribution of energy	Energy access and reliability may be compromised, hindering economic development and social progress
Energy access inequality	Unequal distribution of energy access between urban and rural areas or different socioeconomic groups	Limited energy access exacerbates poverty and hinders social and economic development in marginalized communities
Dependency on traditional fuels	Reliance on traditional and often polluting energy sources, such as biomass and coal	Health issues, environmental degradation, and limited progress toward sustainable development goals
Policy and regulatory gaps	Weak or inconsistent policy frameworks and regulatory environments	Inadequate regulations can deter investments, hamper project development, and hinder the transition to cleaner energy sources
Lack of technical capacity	Limited technical expertise and human resources in the energy sector	Planning, implementing, and maintaining energy projects becomes challenging without skilled personnel, hindering progress
Climate vulnerability	Developing countries are often more vulnerable to climate change impacts, such as extreme weather events	Climate-related challenges can disrupt energy infrastructure and increase the risk of energy insecurity
Political instability	Political instability and governance issues may create uncertainty for investors and disrupt long-term policy continuity	Investment in the energy sector may be deterred, slowing down the transition to more sustainable energy sources
Access to technology	Limited access to advanced technologies and expertise for clean energy development	Adopting and implementing modern technologies for renewable energy may be slow, hindering progress toward cleaner energy

(continued)

Table 2 (continued)

Constraints	Challenges	Impact
Debt and financing challenges	High levels of debt or challenges in securing favorable financing terms	Developing countries may face difficulties in funding large-scale energy projects, particularly those involving clean energy technologies
Social acceptance and awareness	Limited awareness and social acceptance of the benefits of clean energy	Resistance to change and lack of community engagement can impede the adoption of cleaner energy technologies
Land use and resource conflicts	Competition for land and resources between energy projects and other land uses	Conflicts can arise over land rights, potentially delaying or hindering energy projects

Table 3 Best practices

Characteristics	Aspects/Constituents
International collaboration	Collaborate with international organizations, donor agencies, and other countries to access funding, expertise, and technology
Inclusive policy development	Engage a wide range of stakeholders, including local communities, in the policymaking process to ensure inclusivity and social acceptance
Capacity building	Invest in training programs and capacity building to develop the technical skills needed for effective energy planning and implementation
Adaptive policies	Develop policies that are flexible and adaptive to changing circumstances, allowing for iterative adjustments based on evaluation and feedback
Investment incentives	Offer incentives and guarantees to attract private sector investments in energy projects, fostering economic development
Technology transfer	Facilitate the transfer of clean energy technologies through partnerships, collaborations, and knowledge-sharing initiatives
Climate resilience planning	Integrate climate resilience into energy planning to address vulnerabilities and ensure the sustainability of energy infrastructure
Community engagement	Prioritize community engagement and communication to build awareness, address concerns, and promote the social acceptance of clean energy projects
Policy integration	Integrate energy policies with broader development plans to ensure alignment with social, economic, and environmental goals
Regional cooperation	Explore opportunities for regional cooperation on energy projects, which can enhance efficiency, share resources, and strengthen energy security

fuels, these nations face unique challenges and opportunities in navigating a path towards a more sustainable and diversified energy future [64, 65].

Energy Policy in Oil-Rich Countries: The Historical Context

Oil-rich countries have historically built their economies around the extraction, production, and export of oil. The revenue generated from the oil sector has often been a primary driver of economic growth, job creation, and government revenues. Energy policies in these nations have traditionally focused on optimizing oil extraction, maximizing revenue, and ensuring energy security [66, 67].

The dependence on oil exposes these countries to the volatility of global energy markets. Fluctuations in oil prices can have cascading effects on national economies, leading to periods of prosperity or economic downturns. Energy policies have traditionally grappled with the challenge of mitigating these economic vulnerabilities and ensuring long-term stability [68].

Energy Transition: Diversification and Resilience

The global shift towards sustainable energy sources and the growing urgency to address climate change present a compelling case for oil-rich countries to diversify their energy portfolios. Energy transition involves moving away from a sole reliance on fossil fuels and embracing renewable energy sources, such as solar, wind, and nuclear power. This shift is essential for both environmental sustainability and long-term economic resilience [69].

Successful energy transition in oil-rich countries necessitates significant investment in renewable energy infrastructure and the development of innovative technologies. Policies that encourage research and development, incentivize private sector participation, and facilitate the integration of renewable energy into the national grid are crucial for navigating the complexities of transition [70].

Sustainability: Beyond Environmental Considerations

Sustainability in oil-rich countries extends beyond environmental concerns to encompass economic resilience and diversification. Energy policies must be designed to ensure that revenue streams are not solely dependent on oil exports, mitigating the impact of oil price fluctuations on national economies. Economic sustainability involves investing in non-oil sectors, promoting innovation, and fostering a diverse and robust economic landscape [71, 72].

The social dimensions of sustainability are equally vital. Energy policies should prioritize social equity, ensuring that the benefits of economic development and energy access are shared among all segments of the population. Inclusivity in policy formulation and implementation is crucial to prevent the marginalization of certain communities or sectors during the energy transition [31, 73].

Oil-rich countries often wield significant geopolitical influence due to their energy resources. As the global energy landscape evolves, sustainability in these nations requires strategic navigation of geopolitical challenges. Policies must be crafted to maintain diplomatic relationships, foster international collaboration, and align national interests with global efforts to address climate change [74, 75].

Challenges

Diversifying the economy away from oil dependence poses economic challenges, including the need for substantial upfront investments in new industries and the retraining of the workforce. Energy policies must balance the immediate economic benefits of oil extraction with the long-term economic stability offered by diversified sectors [76].

The transition to a sustainable energy future can have social implications, including potential job displacement in the traditional oil sector. Energy policies must incorporate measures to address the social impact of this transition, ensuring the reintegration of affected workers into emerging industries and supporting communities through the change [77, 78].

Political resistance to the transition away from oil can pose a significant obstacle. The influence of powerful stakeholders in the oil industry, coupled with concerns about potential economic upheaval, may lead to policy inertia. Policymakers must navigate these political dynamics, ensuring that energy policies align with long-term sustainability goals [79].

Case Studies

Norway, despite being an oil-rich nation, has successfully leveraged its oil wealth to create a sovereign wealth fund. This fund, fueled by oil revenues, is managed for long-term sustainability and diversification. Norway's experience offers insights into how oil-rich countries can use their resources strategically to ensure economic and financial sustainability beyond the oil era [80, 81].

The United Arab Emirates (UAE) has embarked on a journey of economic diversification through initiatives like Vision 2030. By investing in renewable energy, technology, and other non-oil sectors, the UAE aims to ensure economic sustainability and resilience. Examining the UAE's energy policies provides valuable lessons for other oil-rich nations seeking to navigate the complexities of transition [82, 83]. The UAE launched the Masdar Initiative, which includes the development of Masdar City, a sustainable urban development, and investments in renewable energy projects [84]. The Emirate of Abu Dhabi, part of the UAE, has set ambitious targets for renewable energy, including the Abu Dhabi 2030 Green Energy Plan [85]. The UAE is working on the development of nuclear energy for peaceful purposes, including electricity generation. The Barakah Nuclear Energy Plant in Abu Dhabi is a key project in this regard [86].

NEOM is a flagship project under Vision 2030, aiming to create a futuristic, sustainable city in the northwest of Saudi Arabia (KSA). The project includes plans for a significant portion of renewable energy generation and innovative approaches to urban planning and technology [87]. KSA has shown interest in carbon capture and utilization technologies as a means of mitigating carbon emissions. These technologies aim to capture CO_2 emissions from industrial processes and utilize or store them to prevent their release into the atmosphere [88]. Saudi Arabia has also introduced the concept of a circular carbon economy, emphasizing the importance of reducing, reusing, and recycling carbon emissions. This approach aligns with global efforts to achieve carbon neutrality and reduce the environmental impact of energy production [89, 90]. The Kingdom has shown interest in attracting green investments and financing for sustainable projects. This includes partnerships with international organizations and financial institutions to fund renewable energy initiatives and sustainable development projects [28].

Future Prospects and Strategic Considerations

The future of sustainability in oil-rich countries hinges on technological advancements and innovation. Energy policies must actively encourage research and development in renewable energy, energy storage, and other sustainable technologies. Investing in innovation positions these nations at the forefront of the global transition to a sustainable energy future [63, 91].

Given the global nature of climate change and sustainability challenges, international collaboration is paramount. Energy policies should foster diplomatic ties, encourage partnerships with global organizations, and align with international agreements to collectively address environmental concerns and work towards a sustainable future [35].

The dynamic nature of the energy transition requires policy flexibility. Policymakers must be adaptive and responsive to evolving technological, economic, and environmental landscapes. Policies that incorporate mechanisms for regular review and adjustment ensure that the energy sector remains aligned with long-term sustainability goals [92, 93]. The future of sustainability in oil-rich countries involves a complex and dynamic transition as these nations navigate the global shift towards a more sustainable and low-carbon future. Some major considerations and potential trajectories for oil-rich countries in terms of sustainability are presented in Table 4.

Table 4 Future of sustainability

Factors	Challenge	Future Traits
Diversification of economies	Challenge: Many oil-rich countries heavily depend on oil revenues, making their economies vulnerable to oil price volatility	A sustainable future involves diversifying economies through investments in non-oil sectors such as renewable energy, technology, tourism, and other industries
Transition to renewable energy	Reducing dependence on fossil fuels when the economy is built around oil and gas	Investment in renewable energy sources, such as solar, wind, and hydropower, to shift towards a more sustainable and environmentally friendly energy mix
Investment in technology and innovation	Transitioning from traditional oil and gas sectors to emerging technologies	Focusing on research and development, innovation hubs, and technology transfer to become leaders in clean energy technologies and other high-tech industries
Carbon Capture and Storage (CCS)	Addressing the environmental impact of oil and gas extraction and processing	Implementing and investing in carbon capture and storage technologies to reduce emissions associated with fossil fuel activities
Sustainable resource management	Avoiding over-exploitation and environmental degradation related to resource extraction	Implementing sustainable resource management practices, including environmental impact assessments and responsible extraction techniques
Economic diversification through sovereign wealth funds	Managing sovereign wealth funds in a way that promotes long-term economic stability	Using sovereign wealth funds to invest in diverse, sustainable assets that contribute to economic resilience and development
Green finance and ESG investments	Attracting green investments and aligning with Environmental, Social, and Governance (ESG) criteria	Developing policies and financial instruments that encourage green investments and meet ESG standards
Climate adaptation and resilience	Addressing climate change impacts on vulnerable regions	Investing in infrastructure and policies that enhance resilience to climate change, including water management, agriculture, and urban planning
Corporate Social Responsibility (CSR)	Balancing economic interests with social and environmental responsibilities	Implementing CSR initiatives to benefit local communities, protect the environment, and contribute to sustainable development

(continued)

Table 4 (continued)

Factors	Challenge	Future Traits
Social inclusion and human development	Ensuring that the benefits of economic activities are equitably distributed	Prioritizing social development, education, and healthcare to improve the quality of life for all citizens
International cooperation	Challenge: Balancing national interests with global sustainability goals	Engaging in international collaborations, agreements, and partnerships to address global challenges, such as climate change and environmental conservation
Energy efficiency and conservation	Mitigating energy waste and promoting efficient energy consumption	Implementing policies and initiatives that encourage energy efficiency and conservation practices across sectors
Circular economy practices	Reducing waste and promoting sustainable consumption patterns	Embracing circular economy principles to minimize resource use, reduce waste, and promote recycling and upcycling
Skills development for transition	Ensuring the workforce is equipped with the skills needed for emerging industries	Investing in education and training programs to prepare the workforce for jobs in sustainable sectors
Community engagement and empowerment	Ensuring that local communities have a voice in decision-making processes	Prioritizing community engagement, empowerment, and inclusive development initiatives

Energy Policy Research, Energy Transition, and Sustainability: A Pathway to Achieve UN Sustainable Development Goals

The United Nations Sustainable Development Goals (SDGs) constitute a global agenda for transformative change, aiming to address pressing challenges ranging from poverty and inequality to climate change and environmental degradation. Within this comprehensive framework, energy plays a central role, with specific goals dedicated to ensuring access to affordable, reliable, sustainable, and modern energy for all (SDG 7) [94].

Energy Policy Research: Shaping Informed Decision-Making

Energy policy research serves as the intellectual backbone for shaping effective and informed energy policies. It involves multidisciplinary investigations into the technological, economic, social, and environmental aspects of energy systems. Researchers and scientists contribute valuable insights that policymakers can leverage to design and implement strategies aligning with the broader objectives of sustainable development [95].

Data-driven analysis is fundamental to energy policy research, providing evidence-based insights into current energy consumption patterns, emissions profiles, and the potential of renewable energy sources. Robust data and analytical tools enable policymakers to make informed decisions, prioritize interventions, and measure progress toward SDG 7 targets [96, 97]. Some important major areas of energy research that can contribute to achieving UN SDGs, especially SDG 7 are given herewith:

Access to Modern Energy Services: Research on strategies to increase access to modern energy services in underserved and remote areas, with a focus on off-grid and decentralized renewable energy solutions [98].

Renewable Energy Integration: Investigate policies and frameworks that facilitate the integration of renewable energy sources, such as solar, wind, hydro, and geothermal, into the energy mix to promote sustainability [99].

Energy Efficiency Measures: Explore energy efficiency policies and technologies for households, industries, and transportation to optimize energy use and reduce waste [100].

Clean Cooking Solutions: Research clean and efficient cooking technologies to replace traditional biomass cooking methods, addressing health and environmental issues associated with indoor air pollution [101].

Inclusive Energy Access: Investigate strategies to ensure that vulnerable and marginalized populations, including women and rural communities, have equal access to energy services, fostering inclusivity [102].

Investment and Financing Models: Research innovative financing models and investment strategies to attract funding for energy projects, especially those targeting areas with limited access to capital [103].

Policy and Regulatory Frameworks: Examine the effectiveness of energy policies and regulations in creating an enabling environment for sustainable energy development, including the removal of barriers and incentives for private sector participation [104].

Capacity Building and Skill Development: Explore initiatives to build local capacity and skills in the renewable energy sector, empowering communities to actively participate in the development and maintenance of energy infrastructure [105, 106].

Technology Innovation and Adoption: Investigate the role of technology innovation in expanding energy access, including the use of smart grids, energy storage, and digital solutions to enhance the efficiency and reliability of energy systems [107].

Community Engagement and Social Acceptance: Research approaches to community engagement and the social acceptance of energy projects to ensure that local communities are active participants in, and beneficiaries of, energy initiatives [106, 108].

Monitoring and Evaluation: Develop methodologies for monitoring and evaluating the impact of energy policies on access, affordability, and sustainability, providing insights for continuous improvement [109, 110].

Cross-Sectoral Integration: Explore ways to integrate energy policies with other sectors, such as health, education, and agriculture, to maximize the socio-economic benefits of improved energy access [111].

Resilience to Climate Change: Research strategies to enhance the resilience of energy infrastructure to the impacts of climate change, ensuring the continuity of energy services in the face of environmental challenges [112, 113].

Global Partnerships: Investigate the role of international collaborations, partnerships, and knowledge-sharing in advancing the global agenda for universal energy access [114].

Policy Impact Assessment: Assess the impact of existing energy policies on achieving SDG 7, identifying successful strategies and areas for improvement [115, 116].

The major ways in which data-driven analysis is applied in energy policy research are presented in Table 5.

Energy Transition: A Key Pillar of Sustainable Development

Energy transition involves a fundamental shift from conventional, fossil fuel-based energy systems to cleaner, more sustainable alternatives. This transition is driven by the imperative to mitigate climate change, reduce environmental impact, and ensure a secure and resilient energy future. Aligning energy transition goals with the SDGs is essential to address multiple dimensions of sustainable development concurrently [117].

Renewable energy sources, such as solar, wind, hydro, and geothermal, are at the forefront of the energy transition. Policies promoting the adoption and integration of renewable energy into national grids contribute directly to achieving SDG 7 targets. The decentralization of energy systems and advancements in storage technologies further enhance the viability and sustainability of these sources [118, 119].

Table 5 Data-driven analysis in energy policy research

Description	Objective	Data sources	Analysis technique
Energy consumption and demand analysis	Understand patterns of energy consumption and forecast future demand	Historical consumption data, demographic information, economic indicators	Time-series analysis, regression modeling, machine learning for demand forecasting
Renewable energy potential and integration	Assess the feasibility and potential of renewable energy sources	Geographic information, solar and wind resource data, energy production data	GIS mapping, spatial analysis, optimization models for renewable energy integration
Economic impact assessment	Evaluate the economic impact of energy policies and projects	GDP data, employment statistics, industry-specific economic indicators	Input–output analysis, economic modeling, cost–benefit analysis
Carbon emission and climate change modeling	Analyze greenhouse gas emissions and assess the impact of policies on climate change	Emission inventories, climate data, policy implementation data	Emission modeling, scenario analysis, carbon footprint assessments
Energy efficiency and conservation	Evaluate the effectiveness of energy efficiency programs and policies	Building energy consumption data, efficiency program outcomes	Energy audits, benchmarking, impact evaluations
Smart grids and energy infrastructure analysis	Assess the performance and efficiency of energy infrastructure	Smart meter data, grid operation data, infrastructure investment data	Grid analytics, predictive maintenance models, real-time monitoring
Electric Vehicles (EVs) and transportation analysis	Study the adoption and impact of electric vehicles on energy consumption	EV sales data, charging station locations, transportation patterns	Adoption modeling, transportation demand modeling
Policy effectiveness and impact assessment	Evaluate the effectiveness of energy policies and their impact on key metrics	Policy implementation data, compliance records, outcome indicators	Policy impact assessments, statistical analyses, regression models
International energy trade and cooperation	Analyze energy trade patterns and cooperation between nations	Trade statistics, international agreements, energy prices	Trade flow modeling, geopolitical analysis
Public opinion and social acceptance	Understand public perceptions and acceptance of energy policies	Surveys, social media sentiment analysis, public forums	Sentiment analysis, public opinion surveys, qualitative analysis

(continued)

Table 5 (continued)

Description	Objective	Data sources	Analysis technique
Regulatory compliance and enforcement	Monitor and assess compliance with energy regulations	Regulatory reports, enforcement records, compliance data	Compliance monitoring, enforcement trend analysis
Socioeconomic and environmental justice	Evaluate the distributional impacts of energy policies on different socio-economic groups	Demographic data, environmental impact assessments, community profiles	Equity assessments, vulnerability mapping, justice impact assessments
Scenario planning and future projections	Develop future scenarios to inform long-term energy policy planning	Various data inputs for scenario building, including technological trends, economic projections, and geopolitical factors	Scenario analysis, modeling future trends, sensitivity analysis
Cybersecurity and resilience analysis	Assess the cybersecurity risks and resilience of energy systems	Cybersecurity incident reports, system vulnerability assessments	Threat modeling, cybersecurity risk assessments
Technology adoption and innovation analysis	Study the adoption rates and impact of emerging energy technologies	Technology adoption data, innovation indices, R&D investment data	Technology diffusion models, innovation impact assessments

Sustainability: Integrating Economic, Social, and Environmental Dimensions

Sustainability in the context of energy policy extends beyond environmental considerations to encompass economic and social dimensions. Policies must adopt a holistic approach that addresses the triple bottom line—people, planet, and prosperity. Integrating sustainability principles ensures that energy solutions contribute positively to economic development, social equity, and environmental preservation [120].

Sustainable energy policies prioritize social equity, ensuring universal access to clean energy. Inclusivity considerations involve addressing energy poverty, empowering marginalized communities, and fostering social cohesion. By incorporating these principles, policymakers contribute to the fulfillment of SDG 1 (No Poverty), SDG 5 (Gender Equality), and SDG 10 (Reduced Inequalities) [121]. Sustainability also implies economic resilience and job creation. Policies that stimulate investments in renewable energy projects, energy efficiency initiatives, and sustainable infrastructure contribute to SDG 8 (Decent Work and Economic Growth). These policies not only drive economic development but also establish a foundation for long-term employment opportunities [122].

Environmental stewardship is central to sustainable energy policies. Policies must prioritize the reduction of greenhouse gas emissions, preservation of biodiversity, and the adoption of circular economy principles. In doing so, energy transition becomes an active contributor to SDG 13 (Climate Action) and SDG 15 (Life on Land) [123].

Energy Policy Research as a Catalyst for Achieving SDGs

Energy policy research serves as a catalyst for policy innovation and adaptability. Researchers can identify emerging trends, assess the effectiveness of existing policies, and propose adaptive measures. This iterative process ensures that energy policies remain aligned with the evolving needs of society and the dynamic landscape of sustainable development [124, 125].

Effective energy policies hinge on stakeholder engagement and public awareness. Research can identify stakeholder interests, assess public perceptions, and provide insights into communication strategies. By fostering transparency and inclusivity, policymakers enhance the social acceptability and effectiveness of energy transition initiatives [126, 127].

Global Exemplars in Pursuit of SDGs

Costa Rica stands out as a global exemplar in sustainable energy. With a commitment to carbon neutrality by 2021, Costa Rica's energy mix relies heavily on renewable sources, including hydroelectric, wind, and geothermal. The success of Costa Rica's energy transition demonstrates the feasibility of aligning energy policies with SDGs [128].

Germany's Energiewende, or energy transition, is a comprehensive strategy to shift away from nuclear and fossil fuel-based energy sources. The country's commitment to renewable energy and energy efficiency aligns with SDG 7 while contributing to economic growth, innovation, and climate action [129].

Denmark is a global leader in renewable energy, particularly wind power. The country has set ambitious targets for carbon neutrality and aims to be free of fossil fuels by 2050. Denmark's success in wind energy, including offshore wind farms, serves as an example of sustainable energy transition [130, 131].

Iceland relies heavily on geothermal power for its energy needs. The country harnesses the geothermal heat from its volcanic activity for electricity generation and heating. Iceland's use of geothermal energy is an exemplary model for sustainable and clean energy [132].

The UAE, particularly Dubai and Abu Dhabi, has invested heavily in solar power projects. The Mohammed bin Rashid Al Maktoum Solar Park in Dubai is one of the largest solar parks globally. The UAE's commitment to renewable energy aligns with its goals for sustainable development. [133, 134].

Norway relies extensively on hydropower for electricity generation, and it has one of the highest shares of renewable energy in its energy mix. The country's use of hydropower is a model for sustainable and reliable energy production [135].

China is the world's largest producer of solar panels and a leading investor in renewable energy. The country has made significant progress in expanding its renewable energy capacity, with a focus on solar and wind power [136].

Scotland has set ambitious targets for renewable energy and aims to be a net-zero carbon emitter by 2045. The country has a strong focus on wind power, both onshore and offshore, and its commitment aligns with broader climate and sustainability goals [137].

Sweden has made significant strides in achieving a low-carbon energy system. The country has a high share of renewable energy in its electricity mix and has implemented policies to promote energy efficiency and reduce carbon emissions. Sweden's focus on hydropower, bioenergy, and district heating, along with efforts to enhance energy efficiency in buildings and transportation, exemplify its commitment to sustainable energy [138, 139].

Challenges and Opportunities

Implementing sustainable energy policies faces economic and political challenges. Initial investments in renewable infrastructure may strain national budgets, and political resistance from vested interests in the fossil fuel industry can hinder progress. However, these challenges present opportunities for innovative financing mechanisms and diplomatic efforts to garner political support [140].

The integration of renewable energy into existing infrastructures poses technological and infrastructural hurdles. Energy policy research can help identify solutions, promote technological advancements, and guide policies that facilitate the necessary upgrades to energy grids and storage capacities [141].

Future Prospects and Recommendations

The future of energy policy research lies in harnessing technological advancements. Artificial intelligence, blockchain, and the Internet of Things offer opportunities for optimizing energy systems, enhancing efficiency, and facilitating decentralized energy production. Policies that encourage research and development in these areas will be instrumental in shaping the future energy landscape [142].

The global nature of sustainability challenges necessitates increased collaboration and knowledge sharing among nations. Policymakers can benefit from international partnerships, joint research initiatives, and shared best practices to accelerate progress toward SDGs. Collaborative efforts can foster a collective commitment to sustainability and amplify the impact of individual national initiatives [143, 144].

Conclusion

Energy policy studies in different countries provide a rich tapestry of experiences, challenges, and innovations. From economic powerhouses to emerging economies, the dynamic interplay between energy policy, economic development, and environmental sustainability underscores the need for thoughtful and context-specific approaches. By examining these diverse case studies, policymakers, researchers, and stakeholders can gain valuable insights that contribute to the global discourse on crafting effective and resilient energy policies for the twenty-first century.

The nexus between energy policy, energy transition, and sustainability represents a dynamic and interconnected framework that demands strategic and forward-thinking approaches. As nations grapple with the urgent need to address climate change, reduce carbon emissions, and foster sustainable development, the role of energy policies becomes increasingly pivotal. Through a holistic understanding of this nexus, policymakers, researchers, and stakeholders can collaboratively chart a course towards a more sustainable and resilient energy future—one that harmonizes economic prosperity with environmental stewardship and social equity. Policy analyses guide the formulation of strategies that facilitate a smooth energy transition, promoting sustainable practices. Technological innovations arising from energy transition efforts, in turn, influence policy evolution. Market dynamics shaped by policies impact the competitiveness of sustainable energy solutions. Global collaboration is paramount to address common challenges and share knowledge, technologies, and resources.

The imperative to address climate change, economic resilience, and social equity necessitates a strategic and forward-thinking approach to policy formulation and implementation. As these nations chart their course towards a sustainable energy future, the lessons learned from successful case studies, combined with innovative policy frameworks, will be instrumental in navigating the unique challenges and opportunities presented by the nexus in oil-rich contexts. Through proactive and holistic policy-making, oil-rich countries can transition towards sustainability, contributing to a global effort to secure a resilient and environmentally responsible future. However, the path to a sustainable energy future is not without challenges. The intermittency of renewable energy sources requires innovative storage solutions, and substantial infrastructure investments are necessary for the widespread adoption of green technologies. Ensuring policy stability is crucial to attract investments and foster an environment conducive to energy transition.

By aligning energy policies with the principles of sustainable development, nations can simultaneously address environmental challenges, promote economic prosperity, and enhance social equity. Energy policy research serves as a linchpin in this endeavor, providing the knowledge base and innovative solutions needed to navigate the complexities of the energy transition. As we stand at the intersection of technological advancements, policy innovation, and global cooperation, the choices made today will shape the trajectory of sustainable development for generations to come.

References

1. Lockwood, M., Kuzemko, C., Mitchell, C., & Hoggett, R. (2017). Historical institutionalism and the politics of sustainable energy transitions: A research agenda. *Environment and Planning C: Politics and Space*, 35, 312 - 333. https://doi.org/10.1177/0263774X16660561.
2. Daniels, G., & Rose, M. (1984). Energy and Transport: Historical Perspectives on Policy Issues. *Technology and Culture*, 25, 362. https://doi.org/10.2307/3104747.
3. Landsberg, H. (1989). US energy policy in historical perspective. https://doi.org/10.1016/0301-4207(89)90003-2.
4. Leung, G., Cherp, A., Jewell, J., & Wei, Y. (2014). Securitization of energy supply chains in China. *Applied Energy*, 123, 316-326. https://doi.org/10.1016/J.APENERGY.2013.12.016.
5. Ren, F., & Long, D. (2021). Carbon emission forecasting and scenario analysis in Guangdong Province based on optimized Fast Learning Network. *Journal of Cleaner Production*, 317, 128408. https://doi.org/10.1016/J.JCLEPRO.2021.128408.
6. Li, Y., Lan, S., Ryberg, M., Pérez–Ramírez, J., & Wang, X. (2021). A quantitative roadmap for China towards carbon neutrality in 2060 using methanol and ammonia as energy carriers. *iScience*, 24. https://doi.org/10.1016/j.isci.2021.102513.
7. Åberg, A., Höffken, J., & Lidström, S. (2018). Looking for perspectives! EU energy policy in context. 47–59. https://doi.org/10.1007/978-3-319-99097-2_4.
8. Mehedintu, A., Soava, G., Sterpu, M., & Grecu, E. (2021). Evolution and Forecasting of the Renewable Energy Consumption in the Frame of Sustainable Development: EU vs. Romania. *Sustainability*. https://doi.org/10.3390/su131810327.
9. Aste, N., Adhikari, R., & Pero, C. (2011). Photovoltaic technology for renewable electricity production: Towards net zero energy buildings. *2011 International Conference on Clean Electrical Power (ICCEP)*, 446-450. https://doi.org/10.1109/ICCEP.2011.6036380.
10. Pietzcker, R., Osorio, S., & Rodrigues, R. (2021). Tightening EU ETS targets in line with the European Green Deal: Impacts on the decarbonization of the EU power sector. *Applied Energy*. https://doi.org/10.1016/J.APENERGY.2021.116914.
11. Neininger, K., Slosarek, T., Marx, C., Wohlbrandt, A., Sukumaran, M., Gu, W., Sieglerschmid, G., Böttinger, E., & Kremer, A. (2023). Interoperability for EU DSM: Implementation of CEF building blocks in the Smart4Health project - success stories and lessons learned. *F1000Research*. https://doi.org/10.12688/f1000research.128648.1.
12. J, C., & Majid, M. (2020). Renewable energy for sustainable development in India: current status, future prospects, challenges, employment, and investment opportunities. *Energy, Sustainability and Society*, 10, 1-36. https://doi.org/10.1186/s13705-019-0232-1.
13. Sokołowski, M. (2019). When black meets green: A review of the four pillars of India's energy policy. *Energy Policy*. https://doi.org/10.1016/J.ENPOL.2019.03.051.
14. Mitra, I., Chhatbar, K., Kumar, A., Giridhar, G., Vashistha, R., Meyer, R., & Schwandt, M. (2014). Solmap: Project in India's Solar Resource Assessment. *International Journal of Renewable Energy Development*, 3, 207–216. https://doi.org/10.14710/ijred.3.3.207-216.
15. Verma, V., & Chauhan, P. (2021). An Effective Assessment of Deen Dayal Upadhyay Grameen Kaushalya Yojana (DDU-GKY) in Providing Skill Training Programme for Youth: A Study of Haryana State. *EPRA International Journal of Multidisciplinary Research (IJMR)*. https://doi.org/10.36713/epra7381.
16. Irfan, M., Zhao, Z., Ikram, M., Gilal, N., Li, H., & Rehman, A. (2020). Assessment of India's energy dynamics: Prospects of solar energy. *Journal of Renewable and Sustainable Energy*, 12, 053701. https://doi.org/10.1063/1.5140236.
17. Horner, N., Oliveira, A., Silberglitt, R., Poppe, M., & Rocha, B. (2016). Energy foresight, scenarios and sustainable energy policy in Brazil. *Foresight*, 18, 535-550. https://doi.org/10.1108/FS-06-2015-0035.
18. Scarrow, R. (2018). Fall of ethanol. *Nature Sustainability*, 1, 533. https://doi.org/10.1038/s41893-018-0162-5.

19. Barra, T., Torres, C., Sasso, M., Pereira, V., Neto, E., & Azevedo, D. (2020). Deconvolution of post- and pre-salt petroleum sources in southeastern offshore Brazilian basins. *Organic Geochemistry*, 104146. https://doi.org/10.1016/j.orggeochem.2020.104146.
20. Oliveira, L., Shayani, R., & Oliveira, M. (2013). Proposed business plan for energy efficiency in Brazil. *Energy Policy*, 61, 523-531. https://doi.org/10.1016/J.ENPOL.2013.05.125.
21. Moraes, G., Machado, R., & Silva, E. (2019). Energy efficiency study throughout the industrial operations of a multinational household appliances manufacturer company in Brazil. *Procedia CIRP*. https://doi.org/10.1016/J.PROCIR.2019.01.111.
22. Palma, D., Souza, A., & Mesquita, A. (2020). Funding and Finance Issues Related to The Decommissioning of Brazilian NPPS. *Brazilian Journal of Radiation Sciences*, 8. https://doi.org/10.15392/BJRS.V8I3A.1369.
23. Guimarães, L. (2011). A efetividade do Programa Nacional de universalização do acesso e uso da energia elétrica—Luz para todos, 1, 207–223. https://doi.org/10.5102/RBPP.V1I2.1204.
24. Yamaji, K. (2015). Energy and Environmental Policy in Japan. *Japan Geoscience Union.*
25. Mishchenko, Y. (2021). Conceptual foundations of Japan's modern energy policy: a post-crisis paradigm shift. 46–61. https://doi.org/10.33920/VNE-04-2101-04.
26. Ayoub, N., & Yuji, N. (2012). Governmental intervention approaches to promote renewable energies—Special emphasis on Japanese feed-in tariff. *Energy Policy*, 43, 191-201. https://doi.org/10.1016/J.ENPOL.2011.12.056.
27. Garni, H., Kassem, A., Awasthi, A., Komljenovic, D., & Al-haddad, K. (2016). A multicriteria decision making approach for evaluating renewable power generation sources in Saudi Arabia. *Sustainable Energy Technologies and Assessments*, 16, 137-150. https://doi.org/10.1016/J.SETA.2016.05.006.
28. Amran, Y., Amran, Y., Alyousef, R., & Alabduljabbar, H. (2020). Renewable and sustainable energy production in Saudi Arabia according to Saudi Vision 2030; Current status and future prospects. *Journal of Cleaner Production*, 247, 119602. https://doi.org/10.1016/j.jclepro.2019.119602.
29. Alarjani, A., Modibbo, U., Ali, I., & Sarkar, B. (2021). A new framework for the sustainable development goals of Saudi Arabia. *Journal of King Saud University - Science*, 33, 101477. https://doi.org/10.1016/J.JKSUS.2021.101477.
30. Khan, Z., Murshed, M., Dong, K., & Yang, S. (2021). The roles of export diversification and composite country risks in carbon emissions abatement: evidence from the signatories of the regional comprehensive economic partnership agreement. *Applied Economics*, 53, 4769 - 4787. https://doi.org/10.1080/00036846.2021.1907289.
31. Sgouridis, S., & Csala, D. (2014). A Framework for Defining Sustainable Energy Transitions: Principles, Dynamics, and Implications. *Sustainability*, 6, 1-22. https://doi.org/10.2139/SSRN.2361603.
32. Bompard, E., Botterud, A., Corgnati, S., Huang, T., Jafari, M., Leone, P., Mauro, S., Montesano, G., Papa, C., & Profumo, F. (2020). An electricity triangle for energy transition: Application to Italy. *Applied Energy*. https://doi.org/10.1016/j.apenergy.2020.115525.
33. Bogdanov, D., Ram, M., Aghahosseini, A., Gulagi, A., Oyewo, A., Child, M., Caldera, U., Sadovskaia, K., Farfán, J., Barbosa, L., Fasihi, M., Khalili, S., Traber, T., & Breyer, C. (2021). Low-cost renewable electricity as the key driver of the global energy transition towards sustainability. *Energy*. https://doi.org/10.1016/J.ENERGY.2021.120467.
34. Kubli, M., & Ulli-Beer, S. (2016). Decentralisation dynamics in energy systems: a generic simulation of network effects. *Energy research and social science*, 13, 71–83. https://doi.org/10.1016/J.ERSS.2015.12.015.
35. Tronchin, L., Manfren, M., & Nastasi, B. (2018). Energy efficiency, demand side management and energy storage technologies – A critical analysis of possible paths of integration in the built environment. *Renewable and Sustainable Energy Reviews*. https://doi.org/10.1016/J.RSER.2018.06.060.
36. Griffiths, S. (2019). Energy diplomacy in a time of energy transition. *Energy Strategy Reviews*. https://doi.org/10.1016/J.ESR.2019.100386.

37. Al-Saidi, M. (2020). From Economic to Extrinsic Values of Sustainable Energy: Prestige, Neo-Rentierism, and Geopolitics of the Energy Transition in the Arabian Peninsula. *Energies*. https://doi.org/10.3390/en13215545.
38. Karagiannopoulou, S., Sariannidis, N., Ragazou, K., Passas, I., & Garefalakis, A. (2023). Corporate Social Responsibility: A Business Strategy That Promotes Energy Environmental Transition and Combats Volatility in the Post-Pandemic World. *Energies*. https://doi.org/10.3390/en16031102.
39. Bell, S. (2020). The Renewable Energy Transition Energy Path Divergence, Increasing Returns and Mutually Reinforcing Leads in the State-Market Symbiosis. *New Political Economy*, 25, 57–71. https://doi.org/10.1080/13563467.2018.1562430.
40. Vujanović, M., Wang, Q., Mohsen, M., Duić, N., & Yan, J. (2020). Recent progress in sustainable energy-efficient technologies and environmental impacts on energy systems. *Applied Energy*, 116280. https://doi.org/10.1016/j.apenergy.2020.116280.
41. Tang, T., Shahzad, F., Ahmed, Z., Ahmad, M., & Abbas, S. (2022). Energy transition for meeting ecological goals: Do economic stability, technology, and government stability matter? *Frontiers in Environmental Science*. https://doi.org/10.3389/fenvs.2022.955494.
42. Pellicer-Sifres, V. (2020). Transformative Energy Transition from the bottom-up: exploring the contribution of grassroots innovations in the Spanish context. *Innovation: The European Journal of Social Science Research*, 33, 124 - 139. https://doi.org/10.1080/13511610.2019.1705146.
43. Greene, M. (2018). Socio-technical transitions and dynamics in everyday consumption practice. *Global Environmental Change*. https://doi.org/10.1016/J.GLOENVCHA.2018.05.007.
44. Wiese, F. (2016). Resilience Thinking as an Interdisciplinary Guiding Principle for Energy System Transitions. *Resources*, 5, 30. https://doi.org/10.3390/RESOURCES5040030.
45. Bolwig, S., Bažbauers, G., Klitkou, A., Lund, P., Blumberga, A., Gravelsins, A., & Blumberga, D. (2019). Review of modelling energy transitions pathways with application to energy system flexibility. *Renewable and Sustainable Energy Reviews*. https://doi.org/10.1016/J.RSER.2018.11.019.
46. Petit, V. (2018). Solutions for a Sustainable Transition. *The New World of Utilities*. https://doi.org/10.1007/978-3-030-00187-2_5.
47. Bolton, R., & Foxon, T. (2015). Infrastructure transformation as a socio-technical process — Implications for the governance of energy distribution networks in the UK. *Technological Forecasting and Social Change*, 90, 538–550. https://doi.org/10.1016/J.TECHFORE.2014.02.017.
48. Truffer, B., Schippl, J., & Fleischer, T. (2017). Decentering technology in technology assessment: Prospects for socio-technical transitions in electric mobility in Germany. *Technological Forecasting and Social Change*, 122, 34–48. https://doi.org/10.1016/J.TECHFORE.2017.04.020.
49. Reis, S. (2017). Energiewende: german energy policy in times of green transition, 12, 229–249. https://doi.org/10.21530/CI.V12N3.2017.649.
50. Westholm, E., & Lindahl, K. (2012). The Nordic welfare model providing energy transition? A political geography approach to the EU RES directive. *Energy Policy*, 50, 328–335. https://doi.org/10.1016/J.ENPOL.2012.07.027.
51. Sharma, T., & Balachandra, P. (2018). Will the integration of renewable energy enable sustainable transition of Indian electricity system? *Energy Strategy Reviews*. https://doi.org/10.1016/J.ESR.2018.06.002.
52. Kiesecker, J., Baruch-Mordo, S., Heiner, M., Negandhi, D., Oakleaf, J., Kennedy, C., & Chauhan, P. (2019). Renewable Energy and Land Use in India: A Vision to Facilitate Sustainable Development. *Sustainability*. https://doi.org/10.3390/su12010281.
53. Sahoo, N., Mohapatra, P., Sahoo, B., & Mahanty, B. (2017). Rationality of energy efficiency improvement targets under the PAT scheme in India – A case of thermal power plants. *Energy Economics*, 66, 279–289. https://doi.org/10.1016/J.ENECO.2017.06.004.
54. Palaniswamy, S., S, S., Saravanan, M., & Anand., M. (2022). Social, Economic and Environmental Impact of Electric Vehicles in India. *Journal of Environmental Science and Management*. https://doi.org/10.47125/jesam/2022_1/08.

55. Pelz, S., Chindarkar, N., & Urpelainen, J. (2021). Energy access for marginalized communities: Evidence from rural North India, 2015–2018. *World Development*, 137, 105204. https://doi.org/10.1016/j.worlddev.2020.105204.
56. Sivaram, V. (2017). The Global Warming Wild Card. *Scientific American*, 316 5, 48–53. https://doi.org/10.1038/scientificamerican0517-48.
57. Ahmad, M., Ali, M., & Rahim, N. (2021). Hydrogen energy vision 2060: Hydrogen as energy Carrier in Malaysian primary energy mix – Developing P2G case. *Energy Strategy Reviews*, 35, 100632. https://doi.org/10.1016/J.ESR.2021.100632.
58. Dixon, R., Scheer, R., & Williams, G. (2011). Sustainable energy investments: contributions of the Global Environment Facility. *Mitigation and Adaptation Strategies for Global Change*, 16, 83–102. https://doi.org/10.1007/S11027-010-9253-Y.
59. Du, X. (2015). An in-depth view of climate change: addressing climate change while making a transition on the development mode. *Chinese Journal of Population Resources and Environment*, 13, 93–98. https://doi.org/10.1080/10042857.2015.1017905.
60. Wagner, E., Breuer, A., & Franz, O. (2017). Designetz: a modular concept for the energy transition—from isolated solutions to an efficient energy system of the future. 2017, 2670–2673. https://doi.org/10.1049/OAP-CIRED.2017.0949.
61. (2020). The SDC International Report 2020: Cooperating for Energy Transition. https://doi.org/10.7146/aul.395.
62. Loorbach, D., & VerbongG., G. (2012). Governing the Energy Transition. Reality, Illusion or Necessity. *Blood*. https://doi.org/10.4324/9780203126523.
63. Araújo, K. (2014). The emerging field of energy transitions: Progress, challenges, and opportunities. *Energy research and social science*, 1, 112–121. https://doi.org/10.1016/J.ERSS.2014.03.002.
64. Liu, J., Hull, V., Godfray, H., Tilman, D., Gleick, P., Hoff, H., Pahl-Wostl, C., Xu, Z., Chung, M., Sun, J., & Li, S. (2018). Nexus approaches to global sustainable development. *Nature Sustainability*, 1, 466–476. https://doi.org/10.1038/s41893-018-0135-8.
65. Benson, D., Gain, A., & Rouillard, J. (2015). Water Governance in a Comparative Perspective: From IWRM to a 'Nexus' Approach? *Water alternatives*, 8, 756–773.
66. Mehrara, M. (2007). Energy consumption and economic growth: The case of oil exporting countries. *Energy Policy*, 35, 2939–2945. https://doi.org/10.1016/J.ENPOL.2006.10.018.
67. Vivoda, V. (2009). Diversification of oil import sources and energy security: A key strategy or an elusive objective? *Energy Policy*, 37, 4615–4623. https://doi.org/10.1016/J.ENPOL.2009.06.007.
68. Randall, L. (1986). Energy demand management and pricing in oil-exporting countries: Venezuela and Mexico. *Annual Review of Energy*, 11, 229–251. https://doi.org/10.1146/ANNUREV.EG.11.110186.001305.
69. Tagliapietra, S. (2019). The impact of the global energy transition on MENA oil and gas producers. *Energy Strategy Reviews*. https://doi.org/10.1016/J.ESR.2019.100397.
70. Fattouh, B., & Sen, A. (2020). Economic Diversification in Arab Oil-Exporting Countries in the Context of Peak Oil and the Energy Transition. *When Can Oil Economies Be Deemed Sustainable?* https://doi.org/10.1007/978-981-15-5728-6_5.
71. Fattouh, B., Poudineh, R., & West, R. (2018). The rise of renewables and energy transition: what adaptation strategy for oil companies and oil-exporting countries? *Energy Transitions*, 3, 45–58. https://doi.org/10.26889/9781784671099.
72. Onifade, S., Alola, A., Erdoğan, S., & Acet, H. (2021). Environmental aspect of energy transition and urbanization in the OPEC member states. *Environmental Science and Pollution Research*, 28, 17158–17169. https://doi.org/10.1007/s11356-020-12181-1.
73. Ciplet, D., & Harrison, J. (2020). Transition tensions: mapping conflicts in movements for a just and sustainable transition. *Environmental Politics*, 29, 435–456. https://doi.org/10.1080/09644016.2019.1595883.
74. Rasoulinezhad, E., Taghizadeh-Hesary, F., Sung, J., & Panthamit, N. (2020). Geopolitical Risk and Energy Transition in Russia: Evidence from ARDL Bounds Testing Method. *Sustainability*. https://doi.org/10.3390/su12072689.

75. Blondeel, M., Bradshaw, M., Bridge, G., & Kuzemko, C. (2021). The geopolitics of energy system transformation: A review. *Geography Compass*. https://doi.org/10.1111/gec3.12580.
76. Fuller, J., & Moore, S. (2018). Pedagogy for the Ethical Dimensions of Energy Transitions from Ethiopia to Appalachia. 2, 1–7. https://doi.org/10.1525/CSE.2018.001214.
77. Thaler, P., & Hofmann, B. (2020). The Impossible Energy Trinity: Energy Security, Sustainability, and Sovereignty in Cross-Border Electricity Systems. *SSRN Electronic Journal*. https://doi.org/10.2139/ssrn.3819010.
78. Cantarero, M. (2020). Of renewable energy, energy democracy, and sustainable development: A roadmap to accelerate the energy transition in developing countries. *Energy research and social science*, 70, 101716. https://doi.org/10.1016/j.erss.2020.101716.
79. Newell, P., & Mulvaney, D. (2013). The political economy of the 'just transition'. *The Geographical Journal*, 179, 132–140. https://doi.org/10.1111/GEOJ.12008.
80. Reiche, D. (2010). Sovereign wealth funds as a new instrument of climate protection policy? A case study of Norway as a pioneer of ethical guidelines for investment policy. *Energy*, 35, 3569–3577. https://doi.org/10.1016/J.ENERGY.2010.04.030.
81. Clark, G., & Monk, A. (2010). The Legitimacy and Governance of Norway's Sovereign Wealth Fund: The Ethics of Global Investment. *Environment and Planning A*, 42, 1723–1738. https://doi.org/10.1068/a42441.
82. Sgouridis, S., Griffiths, S., Kennedy, S., Khalid, A., & Zurita, N. (2013). A sustainable energy transition strategy for the United Arab Emirates: Evaluation of options using an Integrated Energy Model. *Energy Strategy Reviews*, 2, 8–18. https://doi.org/10.1016/J.ESR.2013.03.002.
83. Lim, X., Foo, D., & Tan, R. (2018). Pinch analysis for the planning of power generation sector in the United Arab Emirates: A climate-energy-water nexus study. *Journal of Cleaner Production*, 180, 11–19. https://doi.org/10.1016/J.JCLEPRO.2018.01.158.
84. Randeree, K., & Ahmed, N. (2019). The social imperative in sustainable urban development. *Smart and Sustainable Built Environment*. https://doi.org/10.1108/SASBE-11-2017-0064.
85. Naimi, M., & Ali, M. (2021). Comparative Evaluation of Some Energy Policies in Abu Dhabi Using EnergyPlan Program. *Volume 8A: Energy*. https://doi.org/10.1115/imece2021-72694.
86. Almteiri, M., & Kim, J. (2022). Applications of Machine Learning to Consequence Analysis of Hypothetical Accidents at Barakah Nuclear Power Plant Unit 1. *Energies*. https://doi.org/10.3390/en15166048.
87. Alqahtani, N., & Balta-Ozkan, N. (2021). Assessment of Rooftop Solar Power Generation to Meet Residential Loads in the City of Neom, Saudi Arabia. *Energies*. https://doi.org/10.3390/en14133805.
88. Salas, S., Corrales, M., Hoteit, H., Alafifi, A., & Tasianas, A. (2021). Quantifying Uncertainty through 3D Geological Modeling for Carbon Capture Utilization and Storage in the Unayzah Formation in Saudi Arabia. https://doi.org/10.5194/EGUSPHERE-EGU21-12839.
89. Almulhim, A., & Abubakar, I. (2021). Understanding Public Environmental Awareness and Attitudes toward Circular Economy Transition in Saudi Arabia. *Sustainability*. https://doi.org/10.3390/su131810157.
90. Shehri, T., Braun, J., Howarth, N., Lanza, A., & Luomi, M. (2022). Saudi Arabia's Climate Change Policy and the Circular Carbon Economy Approach. *Climate Policy*, 23, 151–167. https://doi.org/10.1080/14693062.2022.2070118.
91. Raven, R. (2007). Niche accumulation and hybridisation strategies in transition processes towards a sustainable energy system: An assessment of differences and pitfalls. *Energy Policy*, 35, 2390–2400. https://doi.org/10.1016/J.ENPOL.2006.09.003.
92. Pagnier, L., & Jacquod, P. (2017). How fast can one overcome the paradox of the energy transition? A physico-economic model for the European power grid. *Energy*. https://doi.org/10.1016/j.energy.2018.05.185.
93. Bettin, S., & Bettin, S. (2020). Electricity infrastructure and innovation in the next phase of energy transition—amendments to the technology innovation system framework. *Review of Evolutionary Political Economy*, 1, 371–395. https://doi.org/10.1007/s43253-020-00021-4.
94. Elavarasan, R., Pugazhendhi, R., Jamal, T., Dyduch, J., Arif, M., Kumar, N., Shafiullah, G., Chopra, S., & Nadarajah, M. (2021). Envisioning the UN Sustainable Development Goals

(SDGs) through the lens of energy sustainability (SDG 7) in the post-COVID-19 world. *Applied Energy*, 116665. https://doi.org/10.1016/J.APENERGY.2021.116665

95. Hyysalo, S., Lukkarinen, J., Kivimaa, P., Lovio, R., Temmes, A., Hildén, M., Marttila, T., Auvinen, K., Perikangas, S., Pyhälammi, A., Peljo, J., Savolainen, K., Hakkarainen, L., Rask, M., Matschoss, K., Huomo, T., Berg, A., & Pantsar, M. (2019). Developing Policy Pathways: Redesigning Transition Arenas for Mid-range Planning. *Sustainability*. https://doi.org/10.3390/SU11030603.
96. Sun, I., & Kim, S. (2017). Energy R&D towards Sustainability: A Panel Analysis of Government Budget for Energy R&D in OECD Countries (1974–2012). *Sustainability*, 9, 617. https://doi.org/10.3390/SU9040617.
97. Wang, H., Ang, B., & Su, B. (2017). Assessing drivers of economy-wide energy use and emissions: IDA versus SDA. *Energy Policy*, 107, 585-599. https://doi.org/10.1016/J.ENPOL.2017.05.034.
98. Li, J., Liu, P., & Li, Z. (2020). Optimal design and techno-economic analysis of a solar-wind-biomass off-grid hybrid power system for remote rural electrification: A case study of west China. *Energy*, 208, 118387. https://doi.org/10.1016/j.energy.2020.118387.
99. Østergaard, P., Duić, N., Noorollahi, Y., Mikulčić, H., & Kalogirou, S. (2020). Sustainable development using renewable energy technology. *Renewable Energy*, 146, 2430–2437. https://doi.org/10.1016/J.RENENE.2019.08.094.
100. Trianni, A., Cagno, E., & Donatis, A. (2014). A framework to characterize energy efficiency measures. *Applied Energy*, 118, 207–220. https://doi.org/10.1016/J.APENERGY.2013.12.042.
101. Rosenthal, J., Balakrishnan, K., Bruce, N., Chambers, D., Graham, J., Jack, D., Kline, L., Masera, O., Mehta, S., Mercado, I., Neta, G., Pattanayak, S., Puzzolo, E., Petach, H., Punturieri, A., Rubinstein, A., Sage, M., Sturke, R., Shankar, A., Sherr, K., Smith, K., & Yadama, G. (2017). Implementation Science to Accelerate Clean Cooking for Public Health. *Environmental Health Perspectives*, 125, A3–A7. https://doi.org/10.1289/EHP1018.
102. Manepalli, K. (2018). A portrait study of microfinance institutions - inclusive growth. *Indian journal of applied research*. https://doi.org/10.36106/ijar/1813056.
103. Hall, B., Moncada-Paternò-Castello, P., Montresor, S., & Vezzani, A. (2016). Financing constraints, R&D investments and innovative performances: new empirical evidence at the firm level for Europe. *Economics of Innovation and New Technology*, 25, 183–196. https://doi.org/10.1080/10438599.2015.1076194.
104. Foxon, T., Gross, R., Chase, A., Howes, J., Arnall, A., & Anderson, D. (2005). UK innovation systems for new and renewable energy technologies: drivers, barriers and systems failures. *Energy Policy*, 33, 2123–2137. https://doi.org/10.1016/J.ENPOL.2004.04.011.
105. Becker, S., Kunze, C., & Vancea, M. (2017). Community energy and social entrepreneurship: Addressing purpose, organisation and embeddedness of renewable energy projects. *Journal of Cleaner Production*, 147, 25–36. https://doi.org/10.1016/J.JCLEPRO.2017.01.048.
106. Koirala, B., Araghi, Y., Kroesen, M., Ghorbani, A., Hakvoort, R., & Herder, P. (2018). Trust, awareness, and independence: Insights from a socio-psychological factor analysis of citizen knowledge and participation in community energy systems. *Energy research and social science*, 38, 33–40. https://doi.org/10.1016/J.ERSS.2018.01.009.
107. Dedrick, J., Venkatesh, M., Stanton, J., Zheng, Y., & Ramnarine-Rieks, A. (2014). Adoption of smart grid technologies by electric utilities: factors influencing organizational innovation in a regulated environment. *Electronic Markets*, 25, 17–29. https://doi.org/10.1007/s12525-014-0166-6.
108. Armeni, C. (2016). Participation in Environmental Decision-making: Reflecting on Planning and Community Benefits for Major Wind Farms. *Journal of Environmental Law*, 28, 415–441. https://doi.org/10.1093/JEL/EQW021.
109. Chou, J., & Ongkowijoyo, C. (2014). Risk-based group decision making regarding renewable energy schemes using a stochastic graphical matrix model. *Automation in Construction*, 37, 98–109. https://doi.org/10.1016/J.AUTCON.2013.10.010.

110. Chapman, A., McLellan, B., & Tezuka, T. (2016). Proposing an evaluation framework for energy policy making incorporating equity: Applications in Australia. *Energy research and social science*, 21, 54–69. https://doi.org/10.1016/J.ERSS.2016.06.021.
111. Orths, A., Anderson, C., Brown, T., Mulhern, J., Pudjianto, D., Ernst, B., O'Malley, M., McCalley, J., & Strbac, G. (2019). Flexibility from Energy Systems Integration: Supporting Synergies Among Sectors. *IEEE Power and Energy Magazine*, 17, 67–78. https://doi.org/10.1109/MPE.2019.2931054.
112. Panteli, M., & Mancarella, P. (2015). The Grid: Stronger, Bigger, Smarter?: Presenting a Conceptual Framework of Power System Resilience. *IEEE Power and Energy Magazine*, 13, 58–66. https://doi.org/10.1109/MPE.2015.2397334.
113. Panteli, M., & Mancarella, P. (2015). Influence of extreme weather and climate change on the resilience of power systems: Impacts and possible mitigation strategies. *Electric Power Systems Research*, 127, 259–270. https://doi.org/10.1016/J.EPSR.2015.06.012.
114. Chaurey, A., Krithika, P., Palit, D., Rakesh, S., & Sovacool, B. (2012). New partnerships and business models for facilitating energy access. *Energy Policy*, 47, 48–55. https://doi.org/10.1016/J.ENPOL.2012.03.031.
115. Rosales-Asensio, E., Puente-Gil, Á., García-Moya, F., Blanes-Peiró, J., & Simón-Martín, M. (2020). Decision-making tools for sustainable planning and conceptual framework for the energy–water–food nexus. *Energy Reports*, 6, 4–15. https://doi.org/10.1016/j.egyr.2020.08.020.
116. Castor, J., Bacha, K., & Nerini, F. (2020). SDGs in action: A novel framework for assessing energy projects against the sustainable development goals. *Energy research and social science*, 68, 101556. https://doi.org/10.1016/j.erss.2020.101556.
117. Aldieri, L., Gatto, A., & Vinci, C. (2021). Evaluation of energy resilience and adaptation policies: An energy efficiency analysis. *Energy Policy*, 157, 112505. https://doi.org/10.1016/J.ENPOL.2021.112505.
118. Deng, Y., Blok, K., & Leun, K. (2012). Transition to a fully sustainable global energy system. *Energy Strategy Reviews*, 1, 109–121. https://doi.org/10.1016/J.ESR.2012.07.003.
119. Hernandez, R., Jordaan, S., Kaldunski, B., & Kumar, N. (2020). Aligning Climate Change and Sustainable Development Goals with an Innovation Systems Roadmap for Renewable Power. 1. https://doi.org/10.3389/frsus.2020.583090.
120. Hahn, T., Pinkse, J., Preuss, L., & Figge, F. (2014). Tensions in Corporate Sustainability: Towards an Integrative Framework. *Journal of Business Ethics*, 127, 297–316. https://doi.org/10.1007/S10551-014-2047-5.
121. Boyer, R., Peterson, N., Arora, P., & Caldwell, K. (2016). Five Approaches to Social Sustainability and an Integrated Way Forward. *Sustainability*, 8, 1–18. https://doi.org/10.3390/SU8090878.
122. Baumgartner, R., & Rauter, R. (2017). Strategic perspectives of corporate sustainability management to develop a sustainable organization. *Journal of Cleaner Production*, 140, 81–92. https://doi.org/10.1016/J.JCLEPRO.2016.04.146.
123. Wang, X., Khurshid, A., Qayyum, S., & Calin, A. (2021). The role of green innovations, environmental policies and carbon taxes in achieving the sustainable development goals of carbon neutrality. *Environmental Science and Pollution Research*, 1–15. https://doi.org/10.1007/s11356-021-16208-z.
124. Santika, W., Anisuzzaman, M., Bahri, P., Shafiullah, G., Rupf, G., & Urmee, T. (2019). From goals to joules: A quantitative approach of interlinkages between energy and the Sustainable Development Goals. *Energy Research & Social Science*. https://doi.org/10.1016/J.ERSS.2018.11.016.
125. Davide, M., Cian, E., & Bernigaud, A. (2019). Building a Framework to Understand the Energy Needs of Adaptation. *Sustainability*. https://doi.org/10.3390/SU11154085.
126. Chen, X., & Musango, J. (2022). A Conceptual Approach to the Stakeholder Mapping of Energy Lab in Poor Urban Settings. *Sustainability*. https://doi.org/10.3390/su14106233.
127. Mascarenhas, K., Peyerl, D., Weber, N., Mouette, D., Cuellar, W., Meneghini, J., & Moretto, E. (2019). Sustainable Development Goals as a Tool to Evaluate Multidimensional Clean

Energy Initiatives. *World Sustainability Series*. https://doi.org/10.1007/978-3-030-26759-9_37.
128. Salas-Mora, V., & Richmond-Navarro, G. (2020). Safety Design of a Hybrid Wind-Solar Energy System for Rural Remote Areas in Costa Rica. *International Journal of Renewable Energy Research*. https://doi.org/10.20508/ijrer.v10i1.10263.g7916.
129. Kemfert, C. (2017). Germany must go back to its low-carbon future. *Nature*, 549, 26-27. https://doi.org/10.1038/549026a.
130. Sovacool, B. (2013). Energy policymaking in Denmark: Implications for global energy security and sustainability. *Energy Policy*, 61, 829–839. https://doi.org/10.1016/J.ENPOL.2013.06.106.
131. Johansen, K. (2021). Blowing in the wind: A brief history of wind energy and wind power technologies in Denmark. *Energy Policy*, 152, 112139. https://doi.org/10.1016/J.ENPOL.2021.112139.
132. Wong, K., & Tan, N. (2014). Feasibility of Using More Geothermal Energy to Generate Electricity. *Journal of Energy Resources Technology-transactions of The Asme*, 137, 041201. https://doi.org/10.1115/1.4028138.
133. Franchini, G., Brumana, G., & Perdichizzi, A. (2019). Monitored performance of the first energy+ autonomous building in Dubai. *Energy and Buildings*, 205, 109545. https://doi.org/10.1016/j.enbuild.2019.109545.
134. Ramachandran, T., Mourad, A., & Hamed, F. (2022). A Review on Solar Energy Utilization and Projects: Development in and around the UAE. *Energies*. https://doi.org/10.3390/en15103754.
135. Graabak, I., Korpås, M., Jaehnert, S., & Belsnes, M. (2019). Balancing future variable wind and solar power production in Central-West Europe with Norwegian hydropower. *Energy*. https://doi.org/10.1016/J.ENERGY.2018.11.068.
136. Zhu, L., Xu, Y., & Pan, Y. (2019). Enabled comparative advantage strategy in China's solar PV development. *Energy Policy*. https://doi.org/10.1016/J.ENPOL.2019.110880.
137. Alcalde, J., Smith, P., Haszeldine, R., & Bond, C. (2018). The potential for implementation of Negative Emission Technologies in Scotland. *International Journal of Greenhouse Gas Control*. https://doi.org/10.1016/J.IJGGC.2018.06.021.
138. Sovacool, B., Kester, J., Rubens, G., & Noel, L. (2018). Expert perceptions of low-carbon transitions: investigating the challenges of electricity decarbonisation in the Nordic region. *Energy*, 148, 1162–1172. https://doi.org/10.1016/J.ENERGY.2018.01.151.
139. Fischer, R., Elfgren, E., & Toffolo, A. (2020). Towards Optimal Sustainable Energy Systems in Nordic Municipalities. *Energies*. https://doi.org/10.3390/en13020290.
140. Mukhi, U., & Quental, C. (2019). Exploring the challenges and opportunities of the United Nations sustainable development goals: a dialogue between a climate scientist and management scholars. *Corporate Governance: The International Journal of Business in Society*. https://doi.org/10.1108/CG-01-2018-0028.
141. Palomares, I., Martínez-Cámara, E., Montes-Soldado, R., García-Moral, P., Chiachío, M., Chiachío, J., Alonso, S., Melero, F., Molina, D., Fernández, B., Moral, C., Marchena, R., Vargas, J., & Herrera, F. (2021). A panoramic view and swot analysis of artificial intelligence for achieving the sustainable development goals by 2030: progress and prospects. *Applied Intelligence*, 1 - 31. https://doi.org/10.1007/s10489-021-02264-y.
142. Mabkhot, M., Ferreira, P., Maffei, A., Podržaj, P., Mądziel, M., Antonelli, D., Lanzetta, M., Barata, J., Boffa, E., Finžgar, M., Paśko, Ł., Minetola, P., Chelli, R., Nikghadam-Hojjati, S., Wang, X., Priarone, P., Lupi, F., Litwin, P., Stadnicka, D., & Lohse, N. (2021). Mapping Industry 4.0 Enabling Technologies into United Nations Sustainability Development Goals. *Sustainability*. https://doi.org/10.3390/SU13052560.
143. Aukes, E., Wilsdon, J., Ordóñez-Matamoros, G., & Kuhlmann, S. (2021). Global resilience through knowledge-based cooperation: a new Protocol for Science Diplomacy. *F1000Research*. https://doi.org/10.12688/f1000research.55199.1.
144. Misiūnaite, R., & Žemaitis, E. (2020). Evaluation of knowledge transfer internationalisation level. https://doi.org/10.3846/bm.2020.643.

SPRINGER NATURE

GPSR Compliance

The European Union's (EU) General Product Safety Regulation (GPSR) is a set of rules that requires consumer products to be safe and our obligations to ensure this.

If you have any concerns about our products, you can contact us on ProductSafety@springernature.com

In case Publisher is established outside the EU, the EU authorized representative is:

Springer Nature Customer Service Center GmbH
Europaplatz 3
69115 Heidelberg, Germany

The manufacturer's authorised representative in the EU is Springer Nature Customer Service Centre GmbH, Europaplatz 3, 69115 Heidelberg, Germany. If you have any concerns regarding our products, please contact ProductSafety@springernature.com

Printed and bound by CPI Group (UK) Ltd, Croydon, CR0 4YY
01/10/2026
02225977-0020